THE FUTURE OF COAL

THE FUTURE OF COAL

Peter James

Second Edition

Palgrave Macmillan

First Edition 1982
Second Edition 1984

Published by
THE MACMILLAN PRESS LTD
London and Basingstoke
Companies and representatives
throughout the world

British Library Cataloguing in Publication Data

James, Peter
The future of coal – 2nd ed.
1. Coal trade
I. Title
333.8'22   HD9540.5

ISBN 978-0-333-36521-2   ISBN 978-1-349-17383-9 (eBook)
DOI 10.1007/978-1-349-17383-9

To Susan and my parents

Contents

List of Figures

List of Tables

Preface to the First Edition

Since the 1973 Yom Kippur War shattered the world's complacency over all resource issues, energy questions have been central to any understanding of global political economy. At first, the world went through an 'oil crisis', predicated on the mistaken assumption that absolute shortages would occur, rather than a painful adjustment of supply and demand patterns to greatly increased real prices. As problems became apparent with other energy sources, the 'energy crisis' came into being and has occupied the headlines until the present day.

Until recently, most thought on the energy crisis has been monolithic in nature. If cheap oil supplies are problematic, the proposed solutions have been to concentrate resources on expanding a limited number of other conventional energy supplies until the 'gap' is closed. In the mid-1970s, nuclear power was to be the saviour of the world but, as its prospects were clouded by the 1979 accident at Harrisburg and other problems, coal has come more to the fore in the latter part of the 1970s and the early 1980s. Unfortunately coal too has its drawbacks, and it will be subject to increasingly critical scrutiny as its expansion and ambitions continue apace.

The problem with such monolithic approaches is an over-dependence on a small number of options. When as is inevitable the problems of implementation stymie the initially grandiose plans, disillusionment and despair results. Fortunately, many energy experts – although fewer governments – have already reached these conclusions, and are now willing to countenance more unorthodox and varied solutions.

My own view is that future generations will be amazed at our reluctance to grasp the obvious about energy matters. The earth itself is a giant energy machine, which can be exploited in many different ways, at many different points, when mankind learns to use it – as is already the case with hydro power. Yet, it is also clear that many such renewable resources provide low-grade or unreliable supplies, and that the world has derived and will continue to derive, incalculable benefit from the high-grade supplies – oil, gas, coal and uranium – on which the modern industrialised world has been founded.

Given that these high-grade supplies are finite resources – compared to the practically infinite nature of renewable energy – the most sensible policy is to reserve the former for premium uses, such as high-temperature requirements or transport fuels, and to use the latter for non-premium uses such as low-temperature space and water heating, with the particular source and technology being used depending on the particular climatic, geographic, and so on, conditions. Such a diverse energy system would be analogous to world food production systems, where a similar mix of high-grade (protein) and low-grade sources (fat, carbohydrate) is used, and with similar climatic and geographical disparities between different regions – rice and beans in Asia, wheat and meat in the West, and so on.

Energy systems represent huge investments of resources, and are not changed overnight. Many decades will pass before such diverse systems could come into being. But, if these are a desirable objective, planning must begin now in order to bring into being. In practice, two questions need to be answered – how can renewable sources be made to meet an increasing share of energy requirements?, and how can inputs to present energy systems be maintained or – in the transition period of coming decades – be increased so that too rapid changes do not cause instability or violence? The first question must be the subject of a different book, but the answer to the second must be a greater use of coal. However, too great an emphasis on coal not only reinforces the 'monolithic' style of energy policy-making – and therefore impedes the growth of renewable energy sources – but also has environmental and other costs. In this book, I have therefore tried to perform a balancing act. Coal expansion will help solve the world's energy problems – but it must not proceed too quickly, and on too great a scale. However, although this is a point of view shared by many observers with more experience and greater knowledge than myself, it has seldom been developed at length. This is one reason why I have long wanted to write a book on this subject.

The second reason for writing this work is the need which I and my former colleagues at the Energy and Planning Group felt for a standard source of information on both the world's coal industries and the technologies they used. In fact, the absence of such a work is due less to the failure of experts in the field to grasp an opportunity, but more to the nature of the coal industry itself. In an area where most production is sold within national boundaries, and where geological conditions, technology, and so on, vary enormously from one country to another there has simply been no need to write an international study. Only in recent years has coal become global in nature, and the logic of such a work become more compelling. This was first recognised in the pioneering World Coal Study of 1980, which gave much valuable information on future coal supply and demand in many countries, and became the 'Bible' of those advocating coal expansion. However, despite its many virtues, the work was primarily compiled by coal producers or users who had an obvious interest in its rosy conclusions. In addition, it excluded or dealt in cursory detail with three of the world's major producers – the USSR, China and South Africa – while many of its assumptions were unfortunately undermined by changes in the world economy between its preparation and publication. Nevertheless, if in the text I have been critical of many of its conclusions and assumptions, it has never been without an awareness that their pioneering work put 'coal on the map' and, in doing so, has made my own task very much easier.

During my work on this book I have been well aware that there are many better qualified than me to write it – and that some have probably taken the wisest course, and avoided the time-consuming and frustrating activity of acquiring data from foreign sources and attempting to master the more intricate details of coal technology. Now that the task is completed, I can at least hope that the book – and its references, which I have tried to make as comprehensive as possible – will provide a useful source to those who are interested in the future of coal.

Preface to the Second Edition

In this second edition I have, wherever possible, updated all statistics to 1981 or later and revised the text to take account of major developments since 1981. A fuller assessment of their significance is contained in a newly written Postscript.

June 1983

Acknowledgements

Many people have helped me in the writing of this book. Of especial value have been my colleagues at the Energy and Planning Group, Nottingham University, particularly Dr Peter Hills, Linda Rosborough, Susan Owens, Barbara Burchell, David Abbott, Frances Baum and John Winward. Many librarians have also dealt with unfailing courtesy with my requests for obscure information, especially Lesley Brown of the National Coal Board Library, and staff at the BP Coal and Commercial Libraries, and the University of Nottingham Library. During the writing of this book I have also had contact with, and information from, many public bodies which are too numerous to cite, but my debts to whom will be apparent in the footnotes. One exception I must make to this anonymity is the Coal Research Unit of the International Energy Agency, particularly its Information Officer, Steve Rogers, and his colleagues, Dr David Hemming, Hugh Lee, Ray Long and John McAdam. Irving Oldfield of the National Coal Board has also been most helpful. These two organisations have also been kind enough to allow me to reproduce figures and tables from their publications, as have the US Department of Energy, the US Central Intelligence Agency, the US Bureau of Mines, the Office of Technology Assessment, the South African Department of Mines, the Canadian Department of Energy, the Australian Division of National Mapping and the World Bank. The World Energy Conference allowed me to use extracts from papers presented at the Munich Conference.

My heartfelt thanks to those above also extend to three other figures, without whom my book would not have been possible. David Cope of the University of Nottingham has been an endless source of support and advice during the long years in which my ideas have been gestating, and has been an invaluable critic of my early drafts. Many of the mistakes I have made were fortunately spotted by his eagle eye. Those which remain are, of course, entirely my own responsibility. Enormous gratitude is also due to my mentor and typist, Susan Waterman, who has borne my alternating bouts of optimism and pessimism about the book with good humour and constant encouragement. Finally, many thanks to my mother for the long hours she too has slaved over a hot typewriter.

Units of Measurement

LENGTH

metre (m); kilometre (km) = 1 000m
1 mile = 1.609km
1 yard = 0.914m
1 foot = 0.305m

AREA

square metre (m^2); hectare (ha) = 10 000m²; square kilometre
(km^2) = 100ha = $1 \times 10^6 m^2$
1 acre = 4 047m² = 0.4047ha
1 square yard = 0.835m²
1 square mile = 2.59km² = 259ha

VOLUME

cubic centimetre (cm^3); litre (l) = 1 000cm³
1 imperial gallon = 4.546l
1 US gallon = 3.785l

MASS (WEIGHT)

kilogramme (kg); tonne = 1 000kg
1 ton (short) = 907kg = 0.907tonne
1 ton (long) = 1 016kg = 1.016tonne

ENERGY, WORK, HEAT

joule (J) and scalar prefixes mega- (MJ), giga- (GJ) and tera- (TJ)
1 British thermal unit (Btu) = 3.6MJ
1 kilowatt hour = 1 055J

POWER

watt (W) and scalar prefixes kilo- (kW), mega- (MW), giga- (GW) and tera- (TW)

1 horsepower = 0.7457kW
1 metric horsepower = 0.7355kW

DECIMAL SCALAR PREFIXES

Abbreviation	Prefix	Scale factor
T	tera	10^{12}
G	giga	10^{9}
M	mega	10^{6}
k	kilo	10^{3}
h	hecto	10^{2}
da	deca	10
d	deci	10^{-1}
c	centi	10^{-2}
m	milli	10^{-3}
μ	micro	10^{-6}
n	nano	10^{-9}
p	pico	10^{-12}

ENERGY EQUIVALENTS (APPROXIMATE)

1 ton of coal equivalent = 1.02 tce
1 ton of oil equivalent = 1.77 tce
1 tonne of oil equivalent = 1.74 tce
1 barrel of oil = 0.23 tce
1 000 cubic feet of natural gas = 0.04 tce
1 000 cubic metres of natural gas = 1.44 tce

Nomenclature and Abbreviations

AC	alternating current
AFB	atmospheric fluidised bed
ASTM	American Society for the Testing of Materials
bbl	barrel
bdoe	barrel(s) per day oil equivalent
bt	billion tonnes (metric)
btce	billion tonnes (metric) coal equivalent
btoe	billion tonnes (metric) oil equivalent
Btu	British thermal unit
CdF	Charbonnages de France
cf	cubic feet
CHP	combined heat and power
CO_2	carbon dioxide
COM	coal-oil mixes
COMECON	Council for Mutual Economic Assistance
CPE	Centrally Planned Economy
DWT	deadweight ton
EEC	European Communities (formerly European Economic Community)
FBC	fluidised bed combustion
FGD	flue gas desulphurisation
GNP	Gross National Product
HCV	high calorific value
IEA	International Energy Agency
kg	kilogram
km	kilometre
kWh	kilowatt hour
lb	pound
LCV	low calorific value
LDCs	Less Developed Countries
mbdoe	million barrels per day oil equivalent

mg	microgram
MJ	Mega-joule
MHD	magnetohydrodynamics
MINOS	Mine Operating System
mst	million short (US) tons
mt	million tonnes (metric)
mtce	million tonnes (metric) coal equivalent
mtoe	million tonnes (metric) oil equivalent
mtpa	million tonnes (metric) per annum
MWe	megawatt(s) of electrical generating capacity = 0.001 GWe = 1000kWe
NCB	National Coal Board
OECD	Organisation for Economic Cooperation and Development
OPEC	Organisation of Petroleum Exporting Countries
PFB	pressurised fluidised bed
ppm	parts per million
psi	pounds per square inch
SASOL	South African Coal, Oil & Gas Company
SNG	Synthetic Natural Gas
SO_2	sulphur dioxide
SRC	solvent refined coal
t	tonnes (metric)
tce	tonnes (metric) coal equivalent
t/d	tonnes (metric) per day
toe	tonnes (metric) oil equivalent
TSP	total suspended particulates
UK	United Kingdom
UNECE	United Nations Economic Commission for Europe
USA	United States of America
USSR	Union of Soviet Socialist Republics
WEC	World Energy Conference
WOCOL	World Coal Study

Note on Statistics

Coal statistics are difficult to interpret. Four different units – tonnes; long tons; short (US) tons and tonnes coal equivalent (tce) – are used for national production figures, sometimes without identification as to which is being used. Similar problems apply to information about national coal reserves and resources, with the additional complication that there is no international standard for distinctions within or between these categories (see Chapter 1 for a fuller discussion). Production figures may also refer to either 'raw' or washed coal which, as there is up to 30 per cent difference in weight between the two categories, can be extremely confusing. Finally, coal statistics do not always take account of the distinction between coal production and consumption, which may be considerable because of the 'buffering' effect of large coal stocks.

In this book, all production figures are for washed coal, and measured in tonnes, unless otherwise stated. They are based primarily on data from the annual international coal survey of *World Coal* magazine, supplemented where necessary by information from the *Mining Annual Review*. Reserve and resource figures are based on those of the World Energy Conference.

S. Schneiderman and R. Strukhoff, 'World Coal Production', *World Coal* 6 (1980) p. 42.
Mining Annual Review (London: Mining Magazine, 1980).
World Energy Conference, *Summary of Energy Resources 1980* (London: World Energy Conference, 1980).

1 Coal: Origins and Exploitation

Coal, like all fossil fuels, is a carbon-rich substance derived from biomass which has grown in the past – in effect, a form of stored solar energy. Unlike petroleum or natural gas, which are thought to originate in the bacterial decay of microscopic marine plants and animals, coal is mainly produced from large plants which grew on land or in shallow swamps.

COAL FORMATION

Coal has been almost continuously created since the beginning of the Carboniferous era, but certain periods have provided more of the tropical warmth and humidity which encourages prolific vegetation growth. Most coals which are exploited today were produced in the Upper and Middle Carboniferous, Permian, Cretaceous and Tertiary periods (see Table 1.1). In general, younger coals have a lower rank than older and are therefore less valuable.

The main requirement for coal formation is a slow subsidence of the earth's surface, allowing a layer of dead plant matter to accumulate to many metres thickness. The processes which follow have been studied in detail for the Carboniferous coal deposits of the United Kingdom (UK), providing a picture which is applicable to other areas and eras.[1] Almost all the present land mass of Great Britain was then a shallow lagoon into which great rivers emptied, with slow subsidence being balanced by sediment deposition. At irregular intervals, the rate of subsidence declined, allowing sediments to fill the lagoon. This was colonised by aquatic plants, followed by trees and giant ferns, thriving in the swamp conditions which persisted for many centuries. Abruptly, the rate of subsidence would increase, causing the water level to rise and the plant layer to be covered by marine or river sediments. This cycle was repeated many times, resulting in many layers of organic material separated by inorganic deposits.

1

TABLE 1.1 Geological age and location of coals

Geological formation	Percentage of total coal resources	Location
Neogene and Paleogene	28.7	Europe, Australia and New Zealand, North America, South America
Cretaceous	16.7	North America, South America, Europe, New Zealand
Jurassic	14.3	Asia, Europe, Australia, North America
Triassic	0.5	Europe, North America,
Permian	24.3	Africa, Antarctica, Australia, Asia, Europe, North America, South America,
Carboniferous	15.6	Europe, North America, Asia

SOURCE World Energy Conference, 'Survey of Energy Resources', (1980).

As organic matter accumulated, aerobic and anaerobic decay took place, reducing its volume by 50–90 per cent and creating peat (diagenesis). During this process the various plant components, for example woody tissue, resin, decomposed in different ways and at different rates, with effects on the chemical and physical properties of the final coal substance. Before complete decomposition occurred, microbial activity was ended by rising concentrations of toxic waste products, and increases in pressure and temperature as the depth of material increased. As this latter phenomenon continued, the chemical and physical properties of the peat changed, turning it into what is in effect an organic sedimentary rock, known as coal (metamorphosis). During this 'coalification' process, the molecular order of the carbon is altered, and moisture is driven off, increasing the coal's calorific value. In the early stages, some plant material is transformed into light organic compounds (volatile matter), which is also expelled in latter stages. Four main stages of coalification are usually distinguished, with each corresponding to a particular class, or rank, of coal, determined by their fixed carbon, volatile matter and moisture contents.

Further sub-divisions are also made (see Table 1.2). The four main coals are:

TABLE 1.2 ASTM classification of coal by rank

VM %[a] FC %[a]	Class	Group	Calorific value[b] Btu per lb	MJ/kg
		Meta-anthracite		
2 98				
	Anthracitic[c]	Anthracite		
8 92				
		Semi-anthracite		
14 86				
		Low volatile bituminous		
22 78				
		Medium volatile bituminous		
31 69				
	Bituminous[d]	High volatile A bituminous		
			14 000	32.6
		High volatile B bituminous		
			13 000	30.2
		High volatile C bituminous		
			11 500	26.7
		Sub-bituminous A[e]		
			10 500	24.4
	Sub-bituminous[f]	Sub-bituminous B		
			9 500	22.1
		Sub-bituminous C		
			8 300	19.3
	Lignitic[f]	Lignite A		
			6 300	14.7
		Lignite B		

[a] Dry, mineral-matter-free basis.
[b] Moist, mineral-matter-free basis.
[c] Non-agglomerating; if agglomerating classified as low volatile bituminous.
[d] Commonly agglomerating.
[e] If agglomerating classified as high volatile C bituminous.
[f] Non-agglomerating.
VM Volatile matter.
FC Fixed carbon.

SOURCE Canadian Department of Energy, 'Coal Resources and Reserves of Canada', (1979).

Lignite A brownish-black substance, usually described as 'brown coal', with untransformed woody matter embedded in decomposed vegetable material. It is liable to spontaneous combustion in air. A high moisture content causes it to disintegrate on drying and gives a low calorific value.

Sub-bituminous A dull-black, waxy, banded substance, in which most woody matter has been transformed. It is intermediate between lignite and full bituminous coals in calorific value and other properties.

Bituminous A dense, black, banded and jointed substance, which breaks into regular blocks. All the original plant material has been transformed, giving a high calorific value. This, plus other properties such as its ignitability and tendency to cake and swell on heating, makes it the most useful and widely produced class of coal.

Anthracite A hard, brittle substance formed under extreme geological conditions. It has a slightly lower calorific value than higher rank bituminous coals, and contains less volatile matter.

The compression which occurs during coalification means that between three and seven metres of compacted plant material are required to produce a one-metre thick bituminous coal seam. The average thickness of all seams is slightly under three metres, although actual thickness vary from a few centimetres to over two hundred metres, as in the Fushun deposits of the North-East China. Similarly, their horizontal extent ranges from a few square metres to thousands of square kilometres, as in the famous Pittsburg bed of the Appalachian USA.

COAL GEOLOGY

About 2 100 sizeable coal deposits have been identified on the earth.[2] Seven of these are giant basins, each containing over 500 bt of coal, of which five – Kansk-Achinsk, Kuznetsk, Lena, Tungusska and Taymyr – are in the USSR. The others are the Appalachian and Alto-Amazona basins of North and South America. Only the Appalachian and Kuznetsk basins have been exploited on a large scale. Four other basins – Donets and Pechora in the USSR, Lower Rhine-Westphalia in West Germany, and the Eastern Interior in the USA – contain between 200 and 500 bt. A further 210 basins have deposits of between 0.5 and 200 bt. The existence of the larger basins is related to the earth's structural features, and they are concentrated in the belts on the periphery of continental platforms. Smaller basins are more unevenly distributed, and many have probably not yet been discovered.

In the hundreds of millions of years since their formation, most of these coalfields have also been affected by tectonic movements with greater or lesser effects on the quality, continuity and mining conditions of coal seams.[3] Both horizontal and vertical stresses can occur in sedimentary rock deposits, resulting in dipping or folding strata, or faulting. As well as rendering coal deposits unworkable by modern technology, these changes can also fracture or cause changes in the coal substance itself. Earth movements are also associated with volcanic activity, and igneous instrusions are a common feature of coalfield geology. These can either cut across the seam or spread in broad sheets along its edges, and their hardness can be a major impediment to mining operations. The heat created by them can also affect the coal, either by increasing its rank or by coking some bituminous deposits. The effect of geological conditions on the material surrounding coal seams can also be an important factor in determining the viability of mining. Often, seams are underlain by clayey material, whose wetness and softness can hamper mining machinery, whilst shale and siltstone roofs are sometimes weak and difficult to support. Where roofs are harder, as with many limestones and sandstones, the stresses of deep mining may make them prone to sudden failure – 'rockburst'. Finally, seams often have been 'washed out' by streams when still at a peat-like stage, with the gaps subsequently filled by sedimentary material which turns to rock and prevents mining when discovered.

In recent decades, considerable advances have been made in coal exploration techniques, allowing the geological conditions of the seam and its surrounding strata to be gauged.[4] Surface drilling and other methods can now provide samples from great depths to measure coal quality and mineral impurities; give information on the dip and configuration of seams and the characteristics of roof and floor rocks; and discover the existence of faults and other obstacles to mining. Recent research has also discovered methods of estimating geological conditions for several hundred metres in advance of working coal faces. Despite these innovations, however, completely accurate geological information is impossible to achieve, and production in all parts of the world is regularly disrupted by the sudden discovery of adverse conditions, particularly in the case of underground mining. In extreme cases, this may necessitate the closure of the site.

COAL COMPOSITION AND PROPERTIES

In general terms, coal is a substance of non-uniform, polymeric structure, the components of which are usually of high molecular weight.[5] Its most important constituent is carbon, which occurs in a large variety of molecules. These are defined in terms of several structural parameters, including size distribution, degree and type of crosslinking, aromaticity, average size of condensed aromatic units, number of hydroxyl groups and scissile bridging structure. During coalification, low molecular weight components, such as methane, phenolic compounds and aliphatic/aromatic hydrocarbons, are produced in the early stages and then progressively driven off. Coal structure also changes greatly. Most low-rank coals have an open structure of randomly oriented small layers, and a large pore system (an average pore diameter of c500Å). High-rank coals have highly orientated and larger layers, an increased aromaticity and a much smaller pore system (an average pore diameter of 5–15Å).

Coal molecules also contain many heteroatoms. Hydrogen occurs in a number of groupings (typically -OH, -CH, $-CH_2$, $-CH_3$, -CO and aromatic -CH), whilst oxygen is usually found in medium molecular weight compounds, especially carbonyl or phenolic hydroxyl groups. Nitrogen and sulphur are also present, and their oxidisation in combustion produces pollutants which are a major impediment to coal use. Most coals have a sulphur level of 0.5–1.5 per cent with the notable exception of those from the Appalachian Basin of the USA, which often contain over 2.5 per cent sulphur.

Coal also contains a varying proportion of mineral matter, which is either bonded to organic molecules or finely divided amongst them. This is mainly in the form of silicon, iron and aluminium oxides, although the oxides or other compounds of most elements are present (see Tables 1.3 and 1.4). Most of this mineral matter must be removed as ash after coal combustion or conversion has occurred, although a small proportion escapes to the atmosphere as fine particulates. The chemical composition of this ash determines the temperatures at which it softens and melts – the 'ash softening point' – which, if low, can make coal use difficult.

The porous nature of most coal also allows large quantities of moisture to be trapped, especially in lower ranks. This inherent moisture reduces the calorific value of coal by consuming heat energy to overcome its latent heat of evaporation.

In addition to its chemical composition, coal can also be analysed by

TABLE 1.3 Ash composition of various coals

	Chemical analyses. % w/w of ash									
	SiO_2	Al_2O_3	Fe_2O_3	TiO_2	P_2O_5	CaO	MgO	Na_2O	K_2O	SO_3
US coals:										
anthracite	47.7–67.7	24.7–43.5	2.1–10.2	1.1–1.8	0.08–3.7	0.2–3.7	0.2–1.2	—	—	0.1–1.1
bituminous	7.1–68.5	4.1–38.9	1.8–43.6	0.5–3.7	0.05–3.1	0.7–36.4	0.1–4.2	0.2–2.8	0.2–3.5	0.1–32.3
subbituminous	16.7–58.3	4.1–35.0	2.7–18.9	0.6–2.3	0.02–3.1	2.2–45.1	0.5–8.0	—	—	2.7–16.1
lignite	6.3–45.2	6.3–22.5	0.9–17.8	0.1–0.8	0.0–1.3	15.3–44.4	3.0–12.2	0.2–11.3	0.1–1.7	6.2–30.3
British coal:										
bituminous	25–50	20–40	0–30	0–3.0	—	1–10	0.5–5.0	1–6	—	1–12
German coals:										
bituminous	25–45	15–21	20–45	—	—	2–4	0.5–1.0	—	—	4.10
brown	7.0–46.3	6.0–29.4	16.6–26.0	—	—	4.1–43.0	0.9–4.0	—	—	2.1–22.0
Australian coal:										
brown	0.1–41.6	0.4–36.9	2.1–29.8	0.1–0.3	—	tr–36.4	0.7–19.5	0.5–6.2	0.1–1.1	8.0–33.1

Source IEA Coal Research, Technical Information Service 'Combustion of low grade coal', (1978).

TABLE 1.4　Concentration of elements in some US coals

Element		Arithmetic mean	Standard deviation	Range of values	
				Min.	Max.
Ag	Silver	0.19 ppm	0.24	< 0.03	0.80
As	Arsenic	14.02 ppm	17.70	0.50	93.00
Au	Gold	0.01 ppm	0.01	< 0.000	0.032
Ba	Barium	130. ppm	150.0	33.0	750.0
Be	Beryllium	1.61 ppm	0.82	0.20	4.00
Ce	Cerium	11. ppm	4.3	4.4	24.0
Cd	Cadmium	2.52 ppm	7.60	0.10	65.00
Co	Cobalt	9.57 ppm	7.26	1.00	43.00
Cr	Chromium	13.75 ppm	7.26	4.00	54.00
Cs	Caesium	1.0 ppm	0.26	0.49	1.5
Cu	Copper	15.16 ppm	8.12	5.00	61.00
Dy	Dysprosium	0.99 ppm	0.30	0.70	1.81
Eu	Europium	0.23 ppm	0.07	0.10	0.40
F	Fluorine	60.94 ppm	20.99	25.00	143.00
Ga	Gallium	3.12 ppm	1.06	1.10	7.50
Ge	Germanium	6.59 ppm	6.71	1.00	43.00
Hf	Hafnium	0.45 ppm	0.14	0.24	0.81
Hg	Mercury	0.20 ppm	0.20	0.02	1.60
I	Iodine	2.0 ppm	1.2	< 1.0	5.8
In	Indium	0.02 ppm	0.02	< 0.008	0.09
La	Lanthanum	6.9 ppm	2.2	3.3	12.0
Lu	Lutetium	0.07 ppm	0.02	0.041	0.11
Mn	Manganese	49.40 ppm	40.15	6.00	181.00
Mo	Molybdenum	7.54 ppm	5.96	1.00	30.00
Ni	Nickel	21.07 ppm	12.35	3.00	80.00
P	Phosphorus	71.10 ppm	72.81	5.00	400.00
Pb	Lead	34.78 ppm	43.69	4.00	218.00
Rb	Rubidium	14. ppm	4.0	7.4	3.6
Sb	Antimony	1.26 ppm	1.32	0.20	8.90
Sc	Scandium	2.4 ppm	0.57	1.4	3.6
Se	Selenium	2.08 ppm	1.10	0.45	7.70
Sm	Samarium	1.1 ppm	0.62	0.4	3.8
Sn	Tin	4.79 ppm	6.15	1.00	51.00
Sr	Strontium	37. ppm	20.0	19.0	130.0
Ta	Tantalum	0.15 ppm	0.05	0.10	0.30
Tb	Turbium	0.15 ppm	0.06	0.04	0.24
Th	Thorium	2.01 ppm	0.47	1.2	3.3
U	Uranium	1.6 ppm	1.1	0.5	4.5
V	Vanadium	32.71 ppm	12.03	11.00	78.00
W	Tungsten	0.74 ppm	0.56	0.04	2.1
Yb	Ytterbium	0.51 ppm	0.13	0.31	0.77
Zu	Zinc	272.29 ppm	694.23	6.00	5350.0

TABLE 1.4 Concentration of elements in some US coals (*Contd*)

Zr	Zirconium	72.46 ppm	57.78	8.00	133.00
*Al	Aluminium	1.29 %	0.45	0.43	3.04
*Ca	Calcium	0.77 %	0.55	0.05	2.67
Cl	Chlorine	0.14 %	0.14	0.01	0.54
*Fe	Iron	1.92 %	0.79	0.34	4.32
*K	Potassium	0.16 %	0.06	0.02	0.43
*Mg	Magnesium	0.05 %	0.04	0.01	0.25
*Na	Sodium	0.05 %	0.04	0.00	0.20
*Si	Silicon	2.49 %	0.80	0.58	6.09
*Ti	Titanium	0.07 %	0.02	0.02	0.15

SOURCE IEA Coal Research, Technical Information Service, 'Trace elements from coal combustion' (1979).

its petrographic nature. This rests on the fact that the parent plant materials retain a separate identity during coalification, and exhibit different properties, although this becomes less marked as the rank of the coal increases.[6] These transformed materials are known as macerals, which can be classified into a small number of maceral groups. All coals can then be considered as rocks which contain varying proportions of macerals, mineral matter and moisture.

In Europe, maceral groups are distinguished by the reflection of light from a thick section, based on the work of Stopes. The three which are usually identified are vitrinite, exinite and inertite. In the USA, four groups are distinguished, on the basis of light transmitted through a thin section, based on the work of Thiessen. These are anthraxylon (corresponding to most vitrinites), translucent attritus (remaining vitrinite and exinite), opaque attritus (some inertite) and fusain (most inertite). Coals composed largely of vitrinite are known as vitrain, while those with a mixture of vitrinite and exinite are known as durain, both these being bright coals. Mixtures of inertite and exinite (durain), or inertite alone (fusain), are known as dull coals. These often appear as separate layers within many coals, accounting for their banded appearance.

Macerals are related to one of coal's most important properties, coking. Reactive macerals – which are mainly vitrinites – when heated become plastic, evolve volatile matter, swell and finally contract and harden. This produces a strong, vesicular and combustible residue – incorporating unchanged unreactive macerals and some ash – known as coke, which is of great importance in iron and steel production.[7] Other coal properties of great importance are its ignitability, which is related to volatile matter content, and its hardness and workability, which affects mining and preparation.[8]

COAL CLASSIFICATION

There is no universally agreed system for classifying coal. Most systems are based on rank, or the degree of metamorphic alteration. This is measured by proximate analysis, which determines moisture, volatile matter, ash and fixed carbon by difference, or ultimate analysis, which determines carbon, hydrogen, sulphur, nitrogen and ash, and estimates oxygen by difference. For pulverised coals, refractivity is also used as an indicator of rank.

For most practical purposes, coals need to be classified quickly, and only a few factors are taken into account.[9] The American ASTM system uses the amount of fixed carbon and the coal's calorific value, calculated on a dry (dmmf) and moist (mmmf) mineral-matter free basis respectively (see Table 1.2). The European UNECE system uses the dry ash-free (daf) volatile matter content, and the moist ash-free (maf) calorific value to distinguish nine coal classes, divided into sub-groups on the basis of their caking properties, with further sub-divisions based on their coking properties. The result is expressed in a three-digit code, with the first indicating class, the second caking behaviour and the third coking behaviour. Other countries have their own systems, as with the UK which uses volatile matter content and coking properties.

Classification of individual coals at present relies on limited and time-consuming sampling, which does not allow variations in coal composition – which can occur within a few metres of the same seam – to be taken into account. In future, continuous nuclear analysis (CONAC) may be used to measure the elements within coal by the gamma-ray frequencies which are emitted during neutron bombardment.[10] This gives instantaneous readings, and can be carried out on a large scale. Moisture content can also be measured by attenuation of microwaves which are passed through the coal.

Classification by rank fails to take account of the fact that coal can follow a number of metamorphic paths, and is not always a good predictor of its technological properties. This has led to classification systems based on their maceral composition, particularly their vitrinite content, which does give such a guide. Eventually, a scientific system which takes account of both rank and petrography is likely to be developed.

COAL RESOURCES

World coal resources are considerable and not yet fully explored. The most comprehensive information is provided by the World Energy Conference, which makes a distinction between resources – the total amount of coal *in situ* – and recoverable reserves – defined as resources which are 'actually recoverable under the technical and economic conditions prevailing today'[11] (see Table 1.5). Clearly, the development of new mining technologies or rising coal prices can transform resources into reserves (see Table 1.6).

The specifications for the World Energy Conference distinctions are, for hard coal (bituminous and anthracite), a maximum depth of 1 500m for recoverable reserves and 2 000m for resources, and a minimum seam thickness of 0.6m for reserves. For brown coal (which is defined, against normal usage, as lignite and sub-bituminous) the specifications reflect their lower value with a maximum depth of 600m for recoverable reserves and 1 500m for resources, and a minimum reserve thickness of 2m.

Unfortunately, some countries do not follow these specifications so that the WEC figures, though the best available, are not always strictly comparable.[12] Thus, the maximum depth for recoverable reserves of hard coal varies from 1 700m in South Korea to 1 500m in the USSR and West Germany and only 300m in the USA, whilst the minimum thickness ranges between 0.3 and 1.5m. The maximum depth of lignite reserves also varies between 600m in West Germany and 30m in the USA. Many countries also have different grades of resources and reserves. The US Geological Survey distinguishes between measured, indicated and inferred reserves and hypothetical or speculative resources, whilst the USSR has five categories of reserve and three of resources. Large quantities of coal are also known to exist below the specified resource depths and may be explored at some point in the future. Finally, detailed research in the USA has shown that existing classifications of reserves and resources are often based on inadequate information, and that some deposits regarded as 'resources' may be mined at a lower cost than others counted as 'measured reserves'.[13]

In 1980, world recoverable reserves of 882 billion tonnes (bt), or 687 billion tonnes coal equivalent (btce), amounted to only 6 per cent of total resources *in situ* of 13 048 bt (11 749 btce).[14]

TABLE 1.5 Recoverable reserves and additional resources of coal (in Gt) according to continents and economic–political groups

Continents or economic–political groups	Proved recoverable reserves						Additional resources in situ					
	Bituminous coal & Anthracite		Sub-bituminous coal		Lignite		Bituminous coal & Anthracite		Sub-bituminous coal		Lignite	
	Gt	%	Gt	%	Gt	%	Gt	%	Gt	%	Gt	%
Africa	32.5	6.7	0.2	0.1	0.0	—	144.4	2.3	1.0	0.0	0.0	—
America	111.4	22.8	96.5	67.5	26.5	10.6	1181.2	19.2	1710.4	44.6	816.9	37.8
Asia	113.9	23.4	1.0	0.7	4.1	1.6	1423.2	23.1	2.8	0.1	59.0	2.7
USSR	104.0	21.3	42.0	29.4	87.0	34.6	2480.0	40.2	2014.0	52.5	1156.0	53.6
Europe	100.5	20.6	1.7	1.2	101.0	40.3	429.5	7.0	1.5	0.0	35.3	1.6
Oceania/ Australia	25.4	5.2	1.6	1.1	32.5	12.9	503.1	8.2	105.5	2.8	92.4	4.3
Total	487.7	100.0	143.0	100.0	251.1	100.0	6161.4	100.0	3835.2	100.0	2159.6	100.0
Common Market	70.0	14.4	0.0	—	35.2	14.0	335.4	5.4	0.0	—	0.0	—
OECD	205.9	42.2	95.6	66.9	98.0	39.0	2007.3	32.6	1794.3	46.8	901.1	41.7
COMECON	134.2	27.5	42.0	29.4	135.7	54.0	2572.3	41.7	2014.0	52.5	1186.5	54.9
Developing countries	22.5	4.6	3.5	2.4	4.1	1.6	214.4	3.5	24.4	0.6	28.2	1.3
OPEC	0.4	0.1	0.3	0.2	0.4	0.2	4.7	0.1	6.0	0.2	17.6	0.8

SOURCE World Energy Conference, 'Survey of Energy Resources', (1980).

TABLE 1.6 Recoverable reserves according to production costs

	Country	< 15	15–30	30–60	> 60
			US$/tonne		
		Bituminous coal & anthracite			
Africa	South Africa	22 090.0	3 200.0		
	Zambia		11.7	3.6	
America	Canada			1 607.0	
	USA	15 627.0	57 614.0	33 942.0	
	Brazil	123.0	66.0		
	Mexico		1 200.0		
	Venezuela		134.0		
Asia	India	12 610.0			
	Indonesia			10.9	
Europe	Belgium	31.0	62.0	62.0	174.0
	France[a]			550.0	
	Federal Republic of Germany			23 991.0	
	Great Britain[a]			45 000.0	
	Netherlands				130.0
	Norway				18.0
	Spain			247.0	151.0
Oceania/ Austr.	Australia[a]	25 400.0			
		Sub-bituminous coal			
America	Canada	2 182.0	neg.		
	USA	91 676.0			
	Argentina			100.0	
	Brazil	600.0	324.0		
	Mexico		384.0		
	Venezuela		2.5	3.8	
Asia	Indonesia		108.4		
	Taiwan			140.0	
Europe	France[a]				10.0
	Spain		123.0		
Oceania/ Austr.	Australia[a]		1 500.0		

TABLE 1.6   Recoverable reserves according to production costs (*Contd*)

			Lignite
America	Canada	2 117.0	
	USA	24 400.0	
Asia	Indonesia		420.0
	Thailand	103.0	
Europe	Federal Republic of Germany	10 000.0	
	Italy	6.0	15.0
	Spain	430.0	
Oceania/ Austr.	Australia[a]	32 440.0	

[a]   Other sources.

SOURCE   World Energy Conference, 'Survey of Energy Resources', (1980).

Nevertheless, these reserves are sufficient to last over two hundred years at present rates of mining and amount to 59 per cent of world recoverable reserves of fossil fuels. Coal also accounts for 84 per cent of the additional resources of all conventional (that is, fossil fuels, plus Uranium-235 fission material) fuel sources.

Both coal reserves and resources are concentrated in a few countries, although this partially reflects a lack of exploration in many parts of the world. Thus, 96 per cent of recoverable reserves of bituminous coal and anthracite are concentrated in nine countries: USA (21.9 per cent), USSR (21.3 per cent), China (20.3 per cent), UK (9.2 per cent), Poland (5.5 per cent), Australia (5.1 per cent), South Africa (5.1 per cent), West Germany (4.9 per cent) and India (2.9 per cent). Sub-bituminous reserves are even more unevenly distributed, with the USSR (64.1 per cent) and USA (29.4 per cent) containing most deposits. Recoverable reserves of lignite are concentrated in the USSR (34.7 per cent), West Germany (14 per cent), Australia (12.9 per cent), East Germany (10 per cent), USA (9.7 per cent) and Yugoslavia (6 per cent). Almost half of the world's additional resources are located in the USSR (see Fig. 1.1).

MINING TECHNOLOGIES

A wide variety of mining techniques are available to exploit coal deposits.[15] The choice of method depends upon geological conditions, safety, productivity, available skills and capital and other factors (see Table 1.7). All share certain characteristics, such as the need to gain access to coal seams through the overlying strata, create a suitable environment for mining to take place and provide a force to break the coal away from surrounding rock (usually by mechanical cutting or explosive blast). In most countries, mining is also highly mechanised, producing improvements in safety and productivity, although at the cost of greater capital requirements and the sterilisation of (underground) reserves which are unsuitable for these methods (see Table 1.8).

Underground Mining

Historically, coal has been won from underground mines, and this is still the main production method in most countries, despite its difficulties.[16] Work must be carried out in confined spaces many metres below the surface, connected to it by a fragile network of tunnels and shafts, which must constantly be protected from crushing by the pressure of overburden. Other problems which require solution are ventilation and the need to remove methane, lighting, drainage, and the transport of men, equipment and coal.

Several types of underground mine are found, according to the way they connect the surface with coal seams. Auger mines are used when a horizontal, or gently dipping, seam occurs near the surface and can be tapped through a short, horizontal tunnel. Their capital and operating costs are low, but most suitable deposits have been exploited. In drift, or slope, mines, an inclined tunnel is used to reach the coal seams, often supplemented with vertical shafts for personnel or emergency access. This has cost advantages over purely shaft mines, and new rock-tunnelling machinery has extended the depths at which it can be used, although geological or hydrological factors restrict its use. Most of the world's underground mines rely on vertical shafts for access, which can be over 1 500m in depth.

Several methods of underground mining are available, with the choice determined by technical characteristics of the coal seam such as thickness, hardness and structural strength, the nature of impurities or irregularities, angle of dip, the amount of methane present, roof and floor quality and hydrological conditions, as well as economic and safety factors.

Until recently, most coal was won by 'room-and-pillar' mining. This involves the driving of parallel tunnels, supported by wooden or metal

TABLE 1.7 Mining methods compared

Mining Method	Coal recovery			Coal production			Seam characteristics[b]								Overburden characteristics					Mine operation				Reclamation effects							Intensive factors		
	High	Medium	Low	High	Medium	Low	Thick	Medium	Thin	Multiple	Pitching	Faults	Gas wells	Hard	Deep	Medium	Shallow	Weak roof	Hard roof	Methane	Water	Safety	Fires	Subsidence	Restoration	Erosion	Acid drainage	Scarring	Spoil	Ground water	Labour	Capital Equipment	Energy
Underground																																	
Room-and-pillar																																	
Conventional			0			0	−	+	+	−	+	+	+	+	+	+	+	−	+	+	+	−	+	−	+	+	−	−	+	−	0		
Continuous			0			0	+	+	−	+	+	+	+	−	+	+	+	−	+	+	+	−	+	−	+	+	−	−	+	−	0		
Longwall	0			0			+	+	−	−	−	−	−	−	+	+	+	+	−	+	−	+	−	+	+	+	−	+	+	−		0	
Shortwall		0			0		−	+	−	+	+	+	+	−	+	+	+	+	+	−	+	+	−	+	+	+	−	+	+	−		0	
Surface																																	
Strip[a]		0		0			+	+	+	+	+	+	+	+	−	+	+	+	+	+	+	+	+	+	−	−	−	−	−	−		0	0
Auger			0		0		+	+	+	−	−	−	+	+	−	+	+	+	+	−	−	+	+	+	−	−	−	−	−	−		0	
Open-pit	0			0			+	−	+	+	+	+	+	+	−	−	+	+	+	+	−	+	+	+	−	−	−	−	−	−		0	0
Quarry-type	0			0			+	−	+	+	+	+	+	+	−	−	+	+	+	+	+	+	+	+	−	−	−	−	−	−		0	0

TABLE 1.7 Mining methods compared (*Contd*)

Advanced																				
Longwall-auger^c	0			+	+	-	+	-	-	+	+	+	+	+	+	-	+	-	+	+	-	-	+	+	+	-	+	+	+	+	0	0
Longwall-caving^d		0		+	-	-	-	-	+	+	+	+	+	+	-	-	-	+	+	+	+	-	-	-	-	+	-	0	0
Borehole^e			0	+	+	+	+	+	+	-	+	+	+	-	+	+	+	+	+	+	+	+	+	-	+	+	+	0
Augering^f	0			+	+	+	+	-	+	+	+	+	+	-	+	+	-	+	+	-	+	-	+	-	+	+	+	0

+ Favourable to indicated mining method.
− Unfavourable to indicated mining method.
0 Characteristic of indicated mining method.

ᵃ The stripping ratio is the cubic yards of overburden removed per ton of coal produced.
ᵇ Coal is economically recoverable if the overburden-to-seam thickness is less than 30:1.
ᶜ A blend of the better characteristics of longwall mining with those of auger mining.
ᵈ Working the bottom of a thick seam and recovering the top coal from the gob.
ᵉ A high-pressure hose through a hole in the surface, washes the coal out, and the slurry is pumped to the surface.
ᶠ Improved augering techniques are to be used for underground mining.

SOURCE United States Department of Energy.

TABLE 1.8 Coal production and mine investment costs in selected coal-producing countries

	Mining tech- nology	Coal type	Mine head production cost Existing mine (US$/ ton-1978)	New mine (US$/ ton-1978)	Incremen- tal mine invest- ment cost (US$/ ton-1978)
Developed markets					
Australia	S	B,C,L	12–15	8–15	30–40
Canada	U	B,C	20–45	n.a.	40–50
	S	S,B,L	6–15	n.a.	20–30
France	U	B	80–95	80–90	n.a.
		L	35–45	n.a.	
Germany, FR	U	B,C		70–100	70–85
		L	10–25		
South Africa	U	B,C	10–12	n.a.	30–35
	S	B,C	8–10	n.a.	n.a.
United Kingdom	U	B,C	45–75	n.a.	70–80
USA	U	B,C	20–30	n.a.	40–55
	S	B,L	8–15	n.a.	10–35
Central planned economies					
China, PR	U	B,C	12–20	n.a.	5–10
Czechoslovakia	U	B,C	30–40	n.a.	60–70
Germany, DR	S	L	8–12	n.a.	15–25
Poland	U	B,C	18–25	n.a.	50–60
	S	L	5–10	n.a.	15–20
USSR	U	B	18–25	n.a.	30–40
	S	L	5–10	n.a.	15–20
Developing countries					
Argentina	U	B	40–45	n.a.	50–60
Brazil	U/S	B	15–25	12–18	25–50
Colombia	U	B,C	5–22	n.a.	n.a.
	S	B	n.a.	25–30	50–60
India	U	B,C	12–25	n.a.	30–35
	S	B,L	20–22	n.a.	n.a.
Indonesia	U	B	35–40	n.a.	n.a.
	S	B	18–20	30–35	50–60
Korea, R.	U	A	20–25	n.a.	35–40
Mexico	U	B,C	15–20	n.a.	45–55
Pakistan	U	B	20–30	n.a.	n.a.
Philippines	U	B	8–21	18–20	30–70
Thailand	S	L	n.a.	7–12	30–35
Venezuela	S	B	n.a.	20–25	50–55
Yugoslavia	U	S,B,L	20–25	n.a.	25–30
	S	S,B,L	11–16	n.a.	10–20

props, from the mine portal. Cross tunnels are dug, creating a chequerboard pattern of coal pillars to support the roof. The deeper the mine, the larger the pillars must be. After all available coal has been extracted in this way, parts of the pillars are mined and the roof allowed to collapse. Until the twentieth century, all room-and-pillar mining was by manual methods, in which the bottom of the seam would be undercut by pickaxe, and holes for explosives drilled in its top. The blast shattered the coal, which was then hand-loaded, usually on to rail cars. Such work required great physical strength and was extremely dangerous. It is largely obsolete in the industrialised world, but is still used in poorer countries such as India, or in poor geological conditions, such as parts of the USSR's Donets Basin.

During the twentieth century, room-and-pillar working has been mechanised, by methods which are now known as conventional mining. In this, coal is extracted in a sequence of operations, each carried out by a separate machine. First, roof support is provided by machine bolting. Undercutting of seams is carried out by self-propelled chain saws, and similar equipment is used to drill holes for blasting, or pumping of compressed air which is safer but slower. The shattered coal is then automatically loaded on to conveyor belts or trucks.

Conventional mining brought great improvements in productivity and safety, but has several disadvantages which led to the development of newer technologies. One problem is the physical difficulty of, and time taken in, moving several pieces of equipment in a confined space. Another is the amount of dust and methane released, which require rock dust spraying and continuous gas monitoring to prevent explosions.

In the USA, conventional room-and-pillar mining has been partially replaced by continuous mining, in which one large machine performs the task previously done by several, with consequent increases in productivity. Several types of continuous miner have been developed, such as boring-, ripping- or milling-machines, but all share the common

---

NOTES TO TABLE 1.8

U = Underground mine.
S = surface mine.
n.a. = not available.
A = anthracite.
B = bituminous.
C = coking coal.
S = sub-bituminous.
L = lignite.

SOURCE World Bank 'Coal Development Potential and Prospects in the Developing Countries', (1979).

characteristic of a rotating head which digs into the seam, with coal pieces transferred by mechanical arms to a conveyor. Usually such machines can cut as much as their length – up to six metres – at which point they require further roof bolting and an extension of ventilation and other support systems. In practice, this, together with conveyor outages, means that continuous mines only operate for 20–30 per cent of shift time.

Room-and-pillar working is only economic at shallow depths, and has low recovery rates. In future, it is likely to be replaced by 'longwall' mining, which is the main method used in Europe and the USSR. In this, parallel development tunnels are dug 100–200m apart, enclosing a large block of coal, from which continuous slices are taken by planer or shearer-loader machines. Hydraulic jacks support the roof as the machine passes and are then self-advanced to provide continuous support at the face, with the roof allowed to collapse behind the machine. Modern supports give two metres of clear space between their base and the face, greatly easing the problems of coal loading and movement, and making ventilation easier. In contrast to this advance longwall method, where cutting moves outward from the mine portal, retreat methods are being increasingly used. Instead of the slow drivage of development tunnels as mining proceeds, long tunnels of up to 1000m are immediately dug. Mining then begins at their end, and proceeds to the mine portal. This has the advantage of lower roof support costs and prior knowledge of seam conditions. However, it involves higher investment costs than advance mining and a longer wait until revenue-earning proceeds.

Longwall mining can be carried out at great depths, or under weak roofs, with good rates of recovery and without the need to abandon equipment. However, longwall mining is more costly than room-and-pillar with the capital cost of opening a 500-foot longwall face in the USA estimated at $4.5 million in 1977.[17] Longwall mining also requires seams of regular cross-section and uniform thickness, which considerably restricts its use. Other problems include the difficulties of keeping development work – which requires drilling through large quantities of hard rock – ahead of mining, and the dependence of production on a small number of units.

To overcome some of these problems, a combination of room-and-pillar and longwall techniques, known as shortwall mining, has been developed in Australia, and is likely to be extensively used in suitable seams in the future.

Surface Mining

Surface mining is generally less costly, more flexible and has higher productivity and recovery rates than underground mining. These latter advantages have been maximised in recent decades by a continuous increase in mine size and the scale of equipment, although the upper limits of this trend may shortly be reached.[18]

The basis of surface mining is the exposure of coal seams by the removal of overlying soil and rocks, followed by mechanical extraction. Usually, topsoil is removed by bulldozer and vehicles, and rock loosened by blasting, with subsequent extraction by draglines, stripping shovels, bucketwheel excavators and bulldozers. Overburden – of which forty tonnes may need to be excavated in order to win one tonne of coal – is usually transferred to worked-out areas for subsequent compaction, replacement of topsoil and revegetation. When exposed, many lignitic and sub-bituminous coals can be scooped up directly but hard coals require similar treatment to the overburden.

Four main methods of surface mining are used. In area mining, a series of long, narrow strips – up to a mile in length and 1000 feet in width – are dug, with draglines transferring debris from the newest to the preceding trench. In open-pit mines, the excavations are larger – up to 2000 feet width – and the overburden is moved from one part of the cut to another. Both these methods are used in relatively flat areas, with open-pit mining being preferred for thick seams.

When coal deposits are found in hilly or mountainous areas, contour mining is practised. Overburden is removed at the point of outcrop, and progressive cuts are made into the slope until the ratio of overburden to coal reaches an uneconomic level. When this point is reached, auger mining – in which a long drill works horizontally into the seam for a distance of up to 200 feet – can be introduced. Where outcrops are near to the top of a hill or mountain, the entire peak may be removed to gain access. Quarry-type mines can also be developed.

New Mining Methods

The cost and danger of present mining technologies, especially underground, have created interest in improvements, or the introduction of new methods.[19] At the coal face, where mechanical breakage has been the normal extraction method, hydraulic mining is likely to increase.[20] This uses water jets to cut and break the coal, which is then transported from the face in a slurry. It is used in the USSR and Canada, where it has given high recovery rates and proved versatile and adaptable, especially in steeply pitching and other disturbed seams. A proposed variant is hydraulic borehole mining, which accesses the seam by a vertical borehole. A water jet at its base breaks up the coal, which is pumped to the surface in a slurry. Although technically feasible, and of obvious value for very steep or vertical seams, economic factors will greatly limit its use in the immediate future.

Other mine equipment is also likely to be improved in the next decade, with particular emphasis on solving the problems of mining at greater depths; unselective cutting; finding methods of tapping multiple, thin and strongly dipping seams; and improving recovery rates, especially from thick seams.[21] Productivity should also be greatly increased by developments in loading and haulage equipment, which are presently the major bottleneck in underground mines.[22]

Safety and economic considerations will also lead to increased underground automation and a reduction in the number of underground workers. In the UK, a Mine Operating System (MINOS) links underground software to a surface computer, which continuously monitors face, transportation and environmental conditions.[23] This will be installed in many collieries during the 1980s. Beyond a certain point of automation, mines may also be given an inert or pressurised atmosphere, with workers equipped with individual life-support systems. However, the unpredictable nature of coal geology and the complexity of even present mining technologies is unlikely to allow fully automated 'mole' or 'telechiric' mining to be practised.

Mining without direct human intervention is more likely to be achieved by *in situ* coal-conversion processes, although these face formidable technical problems. These include: penetration of the coal seams, supplying the extraction media, recovering extraction products, controlling and monitoring underground processes, and overcoming the general problems of high temperature and pressure, likely low reaction rates and the use of hazardous or toxic reagents. Considerable economic and environmental problems must also be faced.

Methods which have been proposed include microbial decomposition, *in situ* combustion (with utilisation of the hot gases), solvent extraction and gasification. The latter appears to be the most promising, although it is only suited to certain types of coal.[24] Several successful experiments have been carried out in Belgium and the USA, producing a low-calorific value gas by the effects of heat and decomposition products from partial coal combustion on the remaining deposits.

COAL PREPARATION

After leaving the face, 'run-of-mine' coal contains many impurities – particularly rock spoil and inorganic sulphur compounds – which restrict its use, and therefore reduce its value. Exhaustion of the best seams, inaccurate cutting by modern machines and more exacting demands by coal consumers have worsened this problem, leading to greater emphasis on preparation or benefication of coal before it leaves the minesite.[25] As coal preparation costs are believed to be offset by savings in transport and power plant operating costs, this trend is likely to continue in the future.

The first stage of preparation is usually size reduction by breakers and rollers, to give a more uniform product and to make subsequent stages easier. However, this increases the amount of small particles, or 'fines', in the coal which can pose problems to some users. After grading, most coals are washed by water. This can remove 10–70 per cent of mineral matter (changing the ash composition and fusion properties), 35 per cent of the coal's sulphur content and upgrade the coal's calorific value by up to 26 per cent. Several cleaning methods are available. Some, such as heavy-media vessels or jigs, use the differences in specific gravity between coal and its impurities, so that one rises and the other sinks. Others, more suited to coal fines, use froth flotation, in which coal particles are caught in a surface froth, created by detergents and reagents, whilst heavier material sinks. A similar process uses agglomerating agents to bind the coal particles together, which then rise to the surface. After cleaning, the coal must be separated from any chemicals used and dewatered, usually by use of vibrating screens, filters or air drying.

Several new methods are likely to be introduced in the future, particularly for the removal of sulphur. One is chemical or microbial leaching of the coal, while separation on the basis of different magnetic properties is claimed to remove all inorganic, and 25–70 per cent of organic, sulphur.

COAL TRANSPORT

Coal is a hard, bulky, dirty solid which, compared to other fossil fuels, is difficult to handle and costly to transport.[26] Many types, especially lignite and sub-bituminous coals, are also prone to disintegration and/or spontaneous combustion after mining and must therefore be used quickly. For these reasons, much coal is used in facilities which are either adjacent to the mine itself or located within a few hundred miles of it. This tendency is likely to increase as the development of high-voltage transmission lines allows power stations to be situated further from their markets, and as large 'coalplexes' producing liquids, gases and chemical feedstocks become an economic proposition. In both cases, transport of the product is likely to be cheaper and easier than moving the coal itself. Despite this, anticipated increases in output will require more physical transport of coal in the future, especially in large countries such as the USSR, USA and China. The costs and difficulties of achieving this may be a major constraint on the future growth of coal production and use (see Fig. 1.2).

Over short distances, much coal is transported by conveyor belt and truck, but the dominant transport mode is railways, often taking the form of fixed coupled, 'unit trains' which constantly circulate from mine to market. The predicted increases in rail-coal traffic in many countries will require costly regrading or new construction of track, although this may be partially offset by the benefits of electrification or improved labour productivity.[27] Although rail will remain the major transport mode until 2000, some coal will transfer to barges which are presently, and are likely to remain, cheaper. However, this trend will be limited by the limited quantity of navigable waterways, and the need for considerable expenditure on upgrading existing, or building new, ones.

One technically-proven, but little-used, transport mode which is expected to be of great importance in the future is slurry pipelines.[28] These pump a suspension of pulverised coal in water (or, in the future, other liquids such as methanol), and should offer cost savings over existing alternatives, particularly where large amounts of coal need to be regularly moved long distances over the same route. However, they suffer from lack of flexibility, opposition from other transport interests, large water requirements and the costs of dewatering the coal at user facilities. Their first uses are likely to be in overcoming existing transport problems in the USA and USSR. At present, only 6 per cent of world coal production is moved by ocean transport, but substantial increases are expected as world coal trade expands and more user facilities are located at coastal sites.[29] The WOCOL Report has projected that coal transport by sea will increase by 4–700 per cent to the year 2000.[30] This

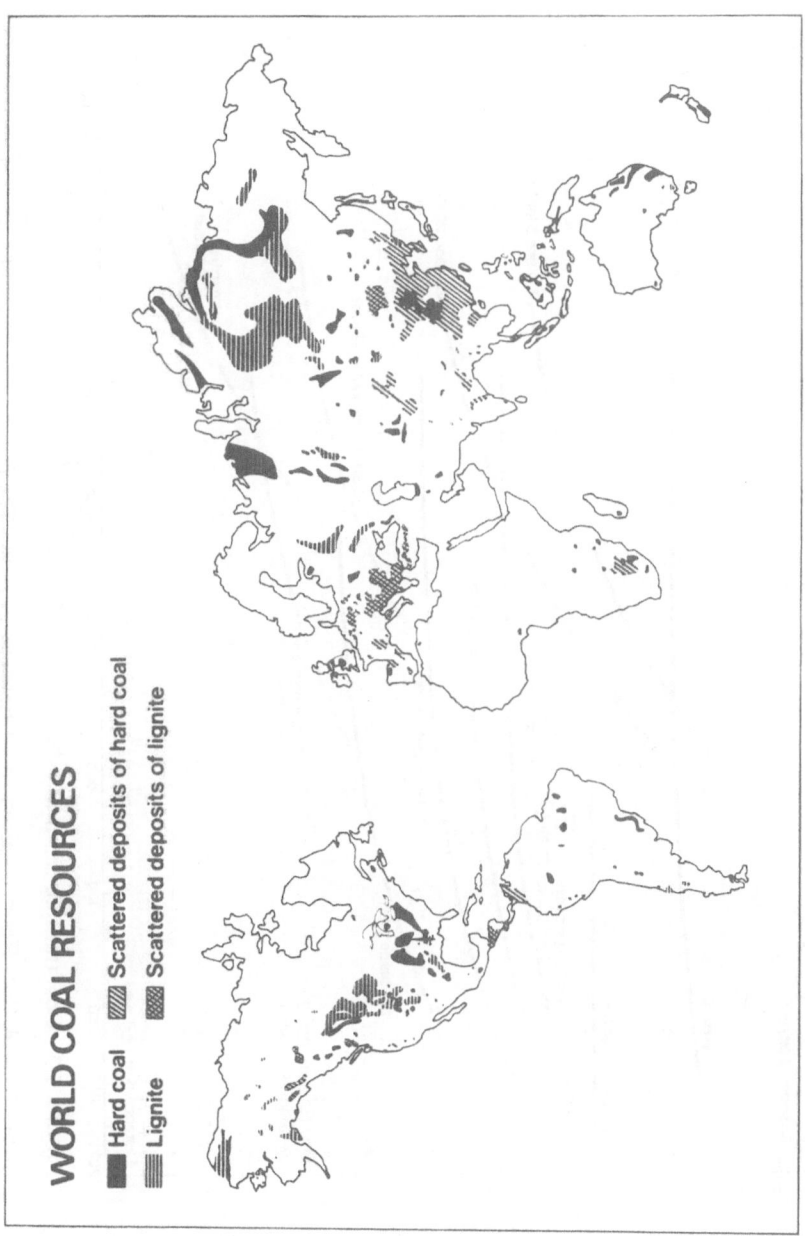

FIGURE 1.1 World coal resources

SOURCE IEA Coal Research

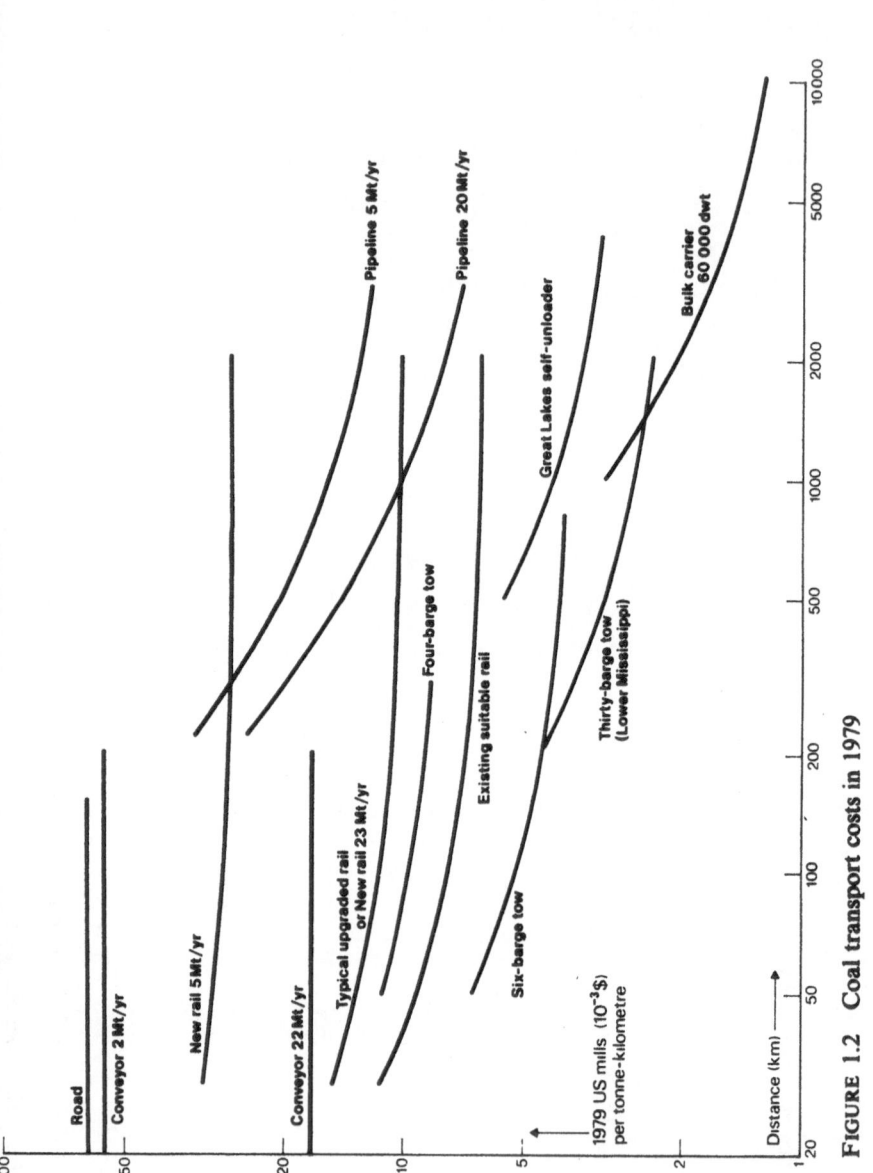

FIGURE 1.2 Coal transport costs in 1979

SOURCE IEA Coal Research, Economic Assessment Service, 'The Future Economics of Coal Transport'. (1980).

will be accompanied by the use of larger carriers, which will change from the typical 50–70 000 DWT bulk carrier of today to specially designed vessels of 100–250 000 DWT by 2000. These will carry the coal to large regional terminals, where the coal will be transhipped to smaller vessels for onward transport to the final users. This will cause substantial falls in freight charges, encouraging coal use further. However, achieving these aims will require substantial investments in ports, ships and other ancillary facilities.

COAL WORKFORCE

The manual technology, and difficult working conditions, of traditional coal mining required a large workforce. Although automation, and declining markets, have greatly reduced their number, mining – particularly underground – is still a substantial employer, and often dominates the labour market in particular regions, such as the Appalachian and Silesian basins of the USA and Poland.[31]

The coal industry requires a variety of workers, including miners (both surface and underground), administrative staff, preparation and other surface plant workers and mine construction personnel. Despite the differences between these occupations, the generally hazardous and ardous nature of mine work, the co-operative working patterns it requires, and a tradition of sons following fathers into the industry has created an unusual degree of collective solidarity amongst the coal workforce. This has provided fertile soil for union activity and allowed miners to undertake longer and more costly strikes than most other industries. These collective tendencies have been reinforced by the physical isolation of many mining communities, and their shared poverty – for the level of miners' wages, although frequently high, has been cyclical in nature.[32]

Many of these traditional characteristics of mineworkers and their lifestyles have been perpetuated in the popular imagination – and have often encouraged opposition to new coal developments because of the social 'problems' they may create. In fact, the conditions of miners and their communities has changed greatly in recent decades. In most countries, they are now near the top of the industrial earnings table and often enjoy other benefits such as early retirement. Their working conditions have also considerably improved, although further progress is possible. The growth of surface, and the automation of underground, mining has required a more skilled and better trained workforce, and has made job movements between coal mining and other industries very much easier. Finally, in many areas, miners and their families have become more integrated into the wider community.

These changes have been most marked in the USA where, according to one study:

> The American coal miner today has little in common with the stereotypes of 30 years ago. The working miner today is far better paid than the miner of the 1940s, in relative as well as absolute terms. Miner's income in 1979 ranks near the top of those of all industrial workers.
>
> Miners today are better educated and better trained than their counterparts of a generation or two ago. They have a vastly improved system of health care. They own new cars, furnish their homes well, take vacations in various parts of the Nation, and in many other ways mirror their colleagues in the auto, steel, or other industries.[33]

Although some observers found this picture overdrawn, the general trend should be maintained in the future. Coal mining will become more capital-intensive, with an increase in surface mining and greater automation underground, requiring a still more skilled workforce and the payment of higher wages or the provision of other benefits. Miners may also place greater emphasis on non-economic factors, such as shorter working hours or the provision of community facilities, and demands for more worker participation could increase. Finally, work-forces will become less homogeneous, with increased recruitment of staff from non-mining backgrounds and the employment of more women and, where applicable, minorities.

The central role of coal in the old industrial economies, and the special nature of mining unionism, has allowed coal workers in many countries to shape the development of national labour, and related political, movements. This social and political power has been increased in the 1970s and early 1980s by the importance of coal in helping many industrialised countries to solve their 'energy crisis', and is often crucial in determining the environment in which the coal industry works.[34] Two prominent examples were the downfall of the Conservative government in Britain as a result of a miners' strike – and its replacement by a Labour government committed to coal expansion – and the role of the Polish miners in the disturbances of 1980. In the short–medium term, at least, this power will remain considerable in many countries and will be an important factor shaping the development of their coal industries.

2 Coal Use: Technology and Markets

The usefulness of coal is due to the special properties of carbon, which is its main constituent. Considerable amounts of the energy which it contains can be used to produce heat and to do work, or to convert the coal substance into other products.[1]

When coal is heated, the complex organic molecules which are its main component begin to decompose, creating smaller and simpler molecules. When this occurs in the presence of oxygen, the familiar reactions of combustion occur. These involve the almost complete oxidisation of the coal and its decomposition products to yield incombustible gases and an inert ash residue. They are exothermic (heat-producing), and have been used as a main source of domestic and industrial heat for many centuries. If oxygen is wholly or partially excluded when coal is heated, the decomposition products can be collected and put to useful purposes. Thus, some bituminous coals which are treated in this way produce a carboniferous, agglomerated residue (coke), which is a vital raw material for the steel industry. These are known as metallurgical, or coking, coals and have a much greater value than other, steam, coals which can only be used to produce heat.

The combustible gases which are produced as a by-product of this pyrolysis can also be purified and piped for later burning at a more convenient time and place. The yields of these gases can be increased by making use of reactions between carbon and gasifying agents, although these are endothermic and require combustion of part of the coal, or an external heat source, to make them go. The gases, as well as the tar and other by-products of pyrolysis, can also be used as chemical feedstocks. Finally, the addition of hydrogen to coal, or the reaction of coal gases with each other, can be used to produce liquid fuels which can substitute for petroleum products.

Although direct combustion will remain the dominant coal use in the present century, it is expected that an increased proportion will be processed into other forms. By the twenty-first century, coal may have returned to its former status of a basic raw material, which serves many different markets.

COAL COMBUSTION

Most of the world's coal production is burnt, and this is likely to remain the case until the next century. This mainly occurs in power stations where heat is converted to electricity. In industrial and residential/commercial markets, however, it is used directly for various purposes.

Conventional Boilers

Several coal combustion technologies exist, with the most suitable depending on the end-use to which the heat will be put, environmental controls and coal properties.[2] The latter includes such factors as sulphur and ash content, caking characteristics, grindability, weatherability and ash-softening characteristics. Although most types of coal can be burnt, higher ranking coals tend to be reserved for coking and other purposes. Despite problems with their low-energy, and high-moisture, content, brown coals can be used with present methods, although new, more efficient techniques are likely to be developed for them.[3]

The most widely-used combustion technology, particularly in power generation, is the pulverised-fuel furnace, which is likely to retain this position. Coal is first ground into fine particles – up to $100\mu m$ diameter – and entrained in pre-heated primary air through burners into the combustion zone. The turbulence created by this high-speed injection, plus that created by rapidly-emitted combustion gases and separately-injected air, ensures that all coal is in contact with oxygen, allowing almost 100 per cent combustion to occur. Flame temperatures can reach $1\,800°C$, and different burner configurations are used to control air and coal inputs to the furnace, maximise heat radiation to the walls and control the formation of nitrous oxides. Most coal ash is removed from beneath the firing zone, although some is entrained in exit gases.

The greatest problem of pulverised-fuel furnaces is that of fouling. A solid layer of slag can coat the furnace walls and heat-exchange tubing, reducing the amount of energy which can be removed and encouraging corrosion. Although air or steam blowers are installed to remove this, manual cleaning is sometimes required, with the resulting furnace 'outage' greatly increasing operating costs. In addition, the high temperatures reached encourage the formation of nitrous oxides. These disadvantages are offset by other factors. The high operating temperature and coal throughput gives a large capacity, while fuel input and heat output can be regulated more easily than in other furnaces. Many different types of coal, or mixtures of coal and other substances, such as refuse, can also be used.

In the cyclone furnace, combustion air is blown into a slightly tilted, horizontal cylinder from several vents. Crushed coal is admitted at the centre and reaches temperatures of up to 1 900°C due to the high degree of turbulence created. At this temperature, coal ash melts and the slag is tapped at the lower end of the cylinder. This design offers a high steam output, low fly-ash formation and reduced costs from the use of crushed, rather than pulverised, coal. Major problems have been experienced with slag viscosity, nitrous oxide formation and high maintenance costs, which have greatly restricted its use. However, it may be introduced to burn coals with ash of low melting point – such as many of the deposits in the western USA – which cause problems in pulverised fuel and other designs.

Traditional furnaces are of fixed-bed design, with the most common varieties being the travelling grate and underfeed stoker. Their common features are the use of crushed coal, which lies on a slowly moving bed, with combustion proceeding as it progresses, leaving an ash residue at its end. They suffer from slow rates of combustion and low heat output, a tendency for coal to cake, and incomplete combustion. However, they have the advantage of relatively low capital and operating costs, and will continue to find a market for industrial coal combustion.

Electricity Generation

In all furnace designs, combustion heat is mainly transferred by radiation to water in the jacketing pipes enclosed in the surrounding walls. In the case of electricity generation, this water boils to form steam, which is then passed through superheater tubes, gaining further heat energy from outgoing combustion gases.[4] The steam then enters a turbine, causing it to spin within a magnetic field and generating AC electricity. In most designs, the steam leaves the turbine about halfway through its passage and is reheated in the boiler before returning to it – a process which improves the energy efficiency of the process. During the last decades, there has been a tendency to increase inlet steam pressures up to 3 500psi and to reheat it up to three times. However, these developments have produced stress problems in critical components, and have created difficulties in the manufacture of the large forgings and castings which were required. Recent units which have been built tend to be 2 400psi and to have a single reheat loop.

Although the efficiency of electricity generation has improved markedly in the past – from 7lb of coal per kWh in 1900 to less than 1lb in 1980 – the thermal efficiency of this route remains low, with only 35–40 per cent of the coal's energy content being captured (less if flue-gas desulphurisation equipment is fitted). Further losses are incurred in the transmission of electricity to its user and in the equipment which finally converts it into useful work.

One way of improving the efficiency of coal-combustion electricity systems is to use the waste heat rejected into turbine cooling water. Such Combined Heat and Power (CHP), or Co-Generation, plants can make use of up to 70 per cent of coal's energy, although with a penalty of slightly reduced electricity output.[5] Markets for hot water can also be difficult to find, with industrial plants and district heating schemes in urban areas the most likely possibilities. Nevertheless, CHP is already well-established in Scandinavia and the USSR, and seems likely to spread in the future.

Fluidised-Bed Combustion

A new combustion technology which holds a great deal of promise for the future is the fluidised bed.[6] In this design, a bed composed of inert ash and limestone or dolomite is fluidised (held in suspension) by the injection of air at its base. Crushed coal is then introduced and – although it forms only a small percentage of the bed (typically 0.5–1 percent) – is quickly burned in the turbulent motion, which carries its combustion heat to all parts of the furnace. This is removed from the bed not only by radiation to the furnace walls and in the exit gases, but also through banks of boiler tubes which are immersed in the bed itself. Despite the relatively low operating temperatures (700–900°C), heat transfer is more efficient than in pulverised-fuel boilers, which reduces the size and cost of the capital equipment. Low temperatures also minimise the formation of nitrous oxides and avoid ash-softening. Spent ash is simply incorporated into the inert bed itself, whilst sulphur emissions are reduced by the reaction of sulphur with the limestone or dolomite to form a dry calcium sulphate, thus avoiding the need for expensive scrubbing equipment.

In the short–medium term, prospects are brightest for beds operating at atmospheric pressure, although pressurised versions – at 4–16 bars – are also being developed. This allows the rate of oxygen supply to be increased, producing a corresponding increase in combustion rate and heat output, so that the size of the furnace can be further reduced. In addition, capture of sulphur dioxide is expected to be more efficient (although only dolomite can be used), whilst nitrous oxide formation should be reduced. After clean-up, the pressurised flue gases can be expanded through a gas turbine, to generate electricity and pressurise incoming combustion air. This combined cycle increases thermal efficiencies to almost 40 per cent (see below). It is estimated that a 500MWe pressurised FBC plant would be only half the size, and considerably cheaper, than a conventional power plant of the same capacity.

In fact, fluidised-bed technology has long been used in the mining, chemical, oil and food industries, although rarely on a large scale or with coal as a fuel. However, despite predictions by the US Department of Energy that 50 per cent of new boiler systems installed in the 1990s will be of fluidised-bed design (mostly operating at atmospheric pressure), many engineering difficulties remain, making it suitable for large-scale electricity generation. One technology assessment has found no insurmountable problems, but identifies several as areas for concern, including solids feeding and distribution systems; system operation and control; materials comparability and corrosion/erosion tendencies; bed performance and combustion efficiency; reliability; safety and environmental problems.[7] These latter include control of particulates, derived from both ash and bed, and the disposal of large volumes of spent sorbet. At present, a 30MWe atmospheric FBC demonstration plant has been operated by Monongahela Power Co. at Rivesville, West Virginia, and the Tennessee Valley Authority (TVA) plans to open a 200MWe unit by the mid-1980s. The British National Coal Board (NCB) also operates a 80MWe pressurised FBC plant for the International Energy Agency (IEA) at Grimethorpe, UK. Full-scale commercialisation of either is unlikely until at least the late 1980s.

MHD

In the long term, electricity may also be generated by Magnetohydrodynamic (MHD) technology.[8] Coal is burnt at a high temperature, and the resulting combustion gases are ionised by the addition of potassium carbonate or another 'seeding' substance. The resulting hot plasma is shot through a channel subjected to a strong magnetic field. The movement of a conductor (the plasma) through this field creates a current which is tapped by electrodes on the inside of the channel.

Although several modes are possible, enthusiasm for MHD is based on the fact that, after passing through the magnetic channel, the gases are hot enough to raise steam for conventional electricity generation. Studies predict that this could result in energy efficiencies of 50 per cent although, due to engineering and material limitations, the first generation of MHD plants is unlikely to exceed 45 per cent efficiency. One study has suggested that if all fossil-fuel plants in Massachusetts had used MHD in 1976, fuel savings would have been almost $300 million.[9]

The technology of MHD also offers some environmental advantages. Sulphur within the coal combines with the potassium carbonate 'seed' to produce potassium sulphate, which can be collected and recycled. Research suggests that sulphur-dioxide emissions would be only 10–20 per cent of current EPA limits, and that particulate emissions would also be reduced – although there may be an increase in sub-micron size particles. Against this, the high combustion temperatures of MHD will produce large quantities of nitrous oxides, and Indian and Japanese scientists have investigated the process as a source of raw material for nitrogenous fertiliser.

Engineering MHD plants will be difficult, due to the high temperature and extreme conditions which they endure. Particular problems are erosion of surfaces by the supersonic gases, with subsequent corrosion, and formation of slag coatings on the electrodes and channel walls. New designs of heat exchanger will also be required, whilst better seed-recovery techniques are necessary. Work is advancing in both the USSR – where two plants with a combined capacity of 25MWe were built in the 1970s, and a 600MWe (gas-fired) demonstration plant is being constructed near to Moscow – and the USA, where a 250MWe plant is planned for Montana in the late 1980s. Interest amongst American utilities is also high, with MHD units seen as a means of converting power plants from oil to coal firing, without installing flue-gas scrubbers. A 50–70MWe MHD unit at the Eriwanda plant of Southern California Edison should be in operation by 1985.

Other Technologies

In the short–medium term, coal–oil mixes (COM) will offer another means of using coal in presently oil-fired boilers.[10] These mixes are slurries of pulverised coal – up to 50 per cent by weight, 35 per cent by heat content – and fuel oil, which can be used in slightly modified oil burners. Coal suspension is maintained by special tanks and paddles and/or chemical additives. Despite some problems, successful trials of COM were held in 1979–80 at the Salem Harbor power plant of New England Power. Utility used is expected to grow substantially in the 1980s, especially in North America.

Shipping companies may also use COM for oil-fired marine boilers, as an interim step before conversion to full coal-firing, which is expected to be widely practised by 2000.[11] Initially, the spreader stoker technology of the last, 1950s, generation of coal-fuelled vessels will be used, but these should eventually be replaced by specially-designed pulverised fuel or fluidised bed boilers. Although the need for coal and ash storage will require a much greater volume than would oil-burning equipment, extra costs should be offset by lower fuel costs. Finally, several non-polluting designs of domestic coal burners have been developed, notably by Britain's National Coal Board, and these will ensure that, on a small scale, coal continues to be used in residential markets.

COAL PROCESSING

Coal in its natural state is a complex substance, with deposits varying in their physical and chemical properties. It is usually contaminated by extraneous matter and, being in solid form, is difficult to handle and transport. Accordingly, there has always been interest in techniques of processing coal into simpler, more uniform or more easily handled products.[12]

Coke-Making

The first method of coal processing to be adopted was its heating in the absence of air (carbonisation), which drives off volatile matter to leave a coherent, cellular mass of carbon, known as coke.[13] In the eighteenth century an English ironmaster, Abraham Derby, substituted this coke for charcoal in iron production and ever since it has been an essential raw material for the iron and steel industry.

Little is known about the properties of coal which determine its coking behaviour, or the mechanisms by which this occurs, and the suitability of individual types must be determined by trial carbonisation. However, the quality of the coke product – which must be strong and resistant to abrasion for metallurgical use – is partially determined by events in the plastic zone (350–500°C), when the coal begins to soften. In practice, good-quality coking coals are of bituminous rank, with a low–medium volatile matter content, and low ash and sulphur levels to avoid contamination in coke-using processes.

Carbonisation is usually carried out in batteries of silica-lined ovens, heated by combustion of the gases which are produced from the coal. They have a typical capacity of 10–20 tonnes and operate on a continuous basis. The coal can be in residence for 12–80 hours, depending on its size and type, and the capacity and temperature of the oven. Most operate at 900–1050°C, although lower temperatures are used when smokeless fuel and non-metallurgical coke is the desired product. When the process if completed, the coke is discharged and quenched in water to prevent further decomposition.

Potential shortages, and the high price of coking coal, has led to the development of new techniques which allow some non-coking coal to be blended with it.[14] These include the preheating of high-volatile bituminous types to reduce their volatile matter content, and form coking, in which crushed coal is formed into briquettes for later carbonisation. As blending techniques continue to improve, the clear division which has hitherto existed between metallurgical and steam coals will become increasingly blurred.

Coke plays three important roles within the blast furnace. Firstly, its partial combustion provides a source of heat. Second, it acts as a source of carbon monoxide which reduces iron ore (which is mainly iron oxides) to its elemental form. Third, it provides physical support for the furnace contents. Good coke must therefore be hard and shatter-proof, so that broken fragments do not impede air flows within the furnace. Until recently, coke has always been indispensable for the two latter purposes, but during the 1960s and 1970s it was partially replaced by oil as a source of heat, causing a marked fall in the amount of coke required for each tonne of steel production. The increased cost of oil is now causing a reversal of this trend, which should offset the increased efficiency of coke use in other areas of the industry for some years to come. In the long run, consumption of coking coal may be reduced by new technical processes, particularly for Directly Reduced Iron (DRI), in which iron ore is considerably purified without the use of a blast furnace.[15] Although coal can be the energy source for this technology, most plants are expected to use cheap natural gas supplies. Demand for coking coal may also be reduced by the growing use of electric arc furnaces, using scrap steel as a raw material, in the industrialised world, although this may be offset by the anticipated expansion of traditional steel-making in developing countries.

Coal Conversion

In general terms, the purpose of coal conversion is to convert a solid, which has a high molecular weight, low volatility and a high carbon/hydrogen ratio, into a liquid or gaseous product which has a low molecular weight, a low carbon/hydrogen ratio and contains no impurities.[16]

For practical purposes, the main difficulties of coal conversion are related to its physical properties. The need for intimate contact between solid carbon surfaces and gaseous or liquid reactants can be hindered by softening and caking tendencies in the coal, which often make (expensive) pre-treatment necessary. Efficient methods must also be found to remove impurities (particularly coal ash and toxic or mutagenic coal products), but these can be complicated by the inertness and 'slagging' (softening and eventual liquefaction) of ash. The presence of large numbers of solid particles, and high temperatures and pressure, can also cause major corrosion problems. Finally, the chemical complexity of liquefaction processes means that reactor engineering must be of a high standard. The flow paths of reactants must be carefully designed, and provision made for efficient exchange of heat between exothermic and endothermic zones.

As, on balance, coal-conversion reactions are endothermic, the need for a heat source – which is provided by combustion of part of the coal itself – is an inherent limitation on the thermal efficiencies of conversion processes. In West Germany and elsewhere, the use of heat from nuclear reactors is under consideration – particularly for gasification – but it seems unlikely that this will be commercialised until the 1990s.[17] Nevertheless, present and projected conversion efficiencies compare favourably with combustion routes for coal use.

GASIFICATION

The first use of coal gas occurred in the eighteenth century, when it lit the Elector's Castle at Dresden.[18] In the UK, a Gas, Light & Coke Company was formed in 1812 and by the late nineteenth century coal gas was used for lighting and heating in most parts of the world. In the twentieth century, low gas yields obtained by simple carbonisation were increased by the use of gasification agents, which combined with unreacted coal (see Fig. 2.1). Four routes are available, using oxygen, steam, carbon dioxide and hydrogen:

$C + \frac{1}{2}O_2 \rightarrow CO$ (exothermic-oxygen limited to avoid complete combustion)
$C + H_2O \rightarrow CO + H_2$ (endothermic)
$C + CO_2 \rightarrow 2CO$ (endothermic)
$C + 2H_2 \rightarrow CH_4$ (exothermic)

Historically, the first two were most important, but technologies have recently been developed which also use the latter two. Since the mid-twentieth century, utilities have turned to oil-derived and natural gas for their supplies, and by the 1970s coal gasification for fuel purposes was almost defunct, although it remained in use for chemical production. Ironically, utilities were closing the last of the old generation of gasifiers at the very time when the energy crisis was renewing interest in their modified successors.

The result of gasification is a mixture of products – typically hydrogen, carbon monoxide, carbon dioxide, hydrogen sulphide, methane and other hydrocarbons – whose proportions vary according to the technology used. After purification, when acid gases (that is, carbon dioxide, hydrogen sulphide) are removed by scrubbing, and excess moisture by dehydrating agents such as sulphuric acid, these are ready for use. If air has been the source of oxygen in the gasifier, the product will be diluted by inert nitrogen and is known as low calorific $(4-8MJ/m^3)$ or 'producer' gas. As transport is uneconomic, this must be used on site, either for industrial purposes or combined cycle electricity generation (see page 45). If more costly elemental oxygen is used, no nitrogen dilution occurs and a medium calorific $(10-16MJ/m^3)$ gas is produced. This can be piped over short distances to serve industrial, domestic or utility markets. It can be upgraded by methanation $(CO + 3H_2 - CH_4 + H_2O)$ to form high calorific $(21 + MJ/m^3)$ gas, also known as Synthetic Natural Gas (SNG).[19] Although this process has a high capital cost and a relatively low thermal efficiency (60–65 per cent), SNG can be distributed along existing natural gas pipelines. As reserves of natural gas are limited, SNG is expected to supplement supplies after the 1990s and to become a major coal market by 2000, particularly in the USA.

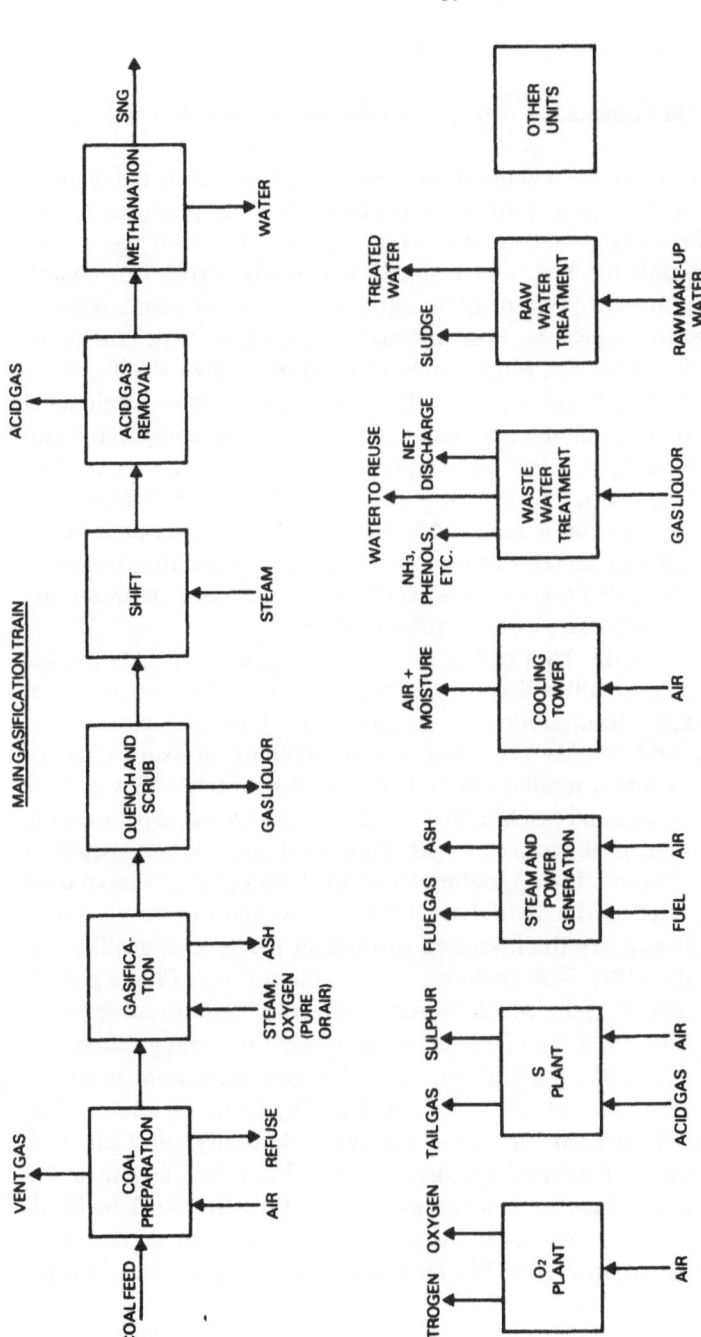

FIGURE 2.1 Flow plan for coal gasification

SOURCE IEA Coal Research, Economic Assessment Service.

Existing Technologies

Three process approaches can be used for gasification (see Fig. 2.2).

Fixed-bed gasifiers are cylinders into which crushed coal is fed from
above. The coal lies on a grate, from beneath which the gasifying agents
are fed. This creates a counter-current system, in which gases flow
upward through the bed, and residual ash slowly settles downward,
eventually falling through the grate into a hopper for disposal. A series
of temperature zones is thus formed, creating a wide range of
decomposition products, which can be difficult to remove. Problems are
also caused by highly caking coals, the solutions to which – mechanical
stirring or pretreatment – add greatly to equipment complexity and
cost. Coal throughput is low and output is further limited by the need to
keep operating temperatures below the ash fusion point, achieved by
steam moderation. This is achieved by the addition of large quantities of
excess steam, which increases costs considerably. Against this, however,
must be set the advantages of relatively simple operation, without any
problems of solids carry-over in the exit gases.

Lurgi is the main fixed-bed design and, despite the disadvantages
cited, gives a thermally efficient and economically viable performance.
One advantage is its operation at pressure (25 bar), which increases coal
throughput and avoids the need for subsequent pressurisation of
product gas. Present models have a capacity several times that of their
1950s predecessors and can be either air- or oxygen-blown. Operations in
the latter mode produce a gas with high methane content, which is
particularly suitable for upgrading to SNG. Lurgi gasifiers have been
successfully used in the SASOL coal conversion plant in South Africa
(see page 49) and are the favoured equipment for several gasification
schemes in the USA. This includes the $3 billion Great Plains project
at Beulah, North Dakota which should produce 137 million cubic feet of
SNG per day by 1986. The favourable prospects for Lurgi gasifiers will
be further enhanced by a high-pressure (100 bar) development of the
present design which is being commercialised by Lurgi-Gesellschaften,
Ruhrgas and Ruhrkohle at a Dorsten, West Germany, pilot plant. A
Wellman-Galusha fixed-bed gasifier has also been installed in a US
brickworks to produce low calorific gas from anthracite. Work on fixed-
bed gasification has also been carried out in Czechoslovakia, where
three plants are in operation. The largest, at Vresová, processes 5000 t/d
of lignite.

In *fluidised-bed* gasifiers fine granules (measuring 1–5mm) of coal are suspended in the upward flow of gasification agents. As with the fluidised-bed boiler, this gives excellent heat and mass transfer characteristics, and operational flexibility. However, temperatures are limited by the need to avoid ash softening and solidification of the bed. Materials handling and solid carry-over problems are also considerable, although the latter can be solved by recycling.

The *Winkler* fluidised-bed gasifier has been available since the 1920s, but is now used only as a source of synthesis gas for fertiliser production. Present models are limited by their operation at atmospheric pressure and can only be used with lignite and other reactive coals. However, the process owners, Davy-Power Gas, are developing higher-temperature pressurised variants which will be able to use less reactive coals. This HTW process is being tested by Phillips Petroleum at a 173t/d US pilot plant, and by Rheinische Braunkohlenwerke at a 25t/d plant near Cologne, West Germany. The latter company is also planning a 2mtpa commercial scale plant at Huerth, West Germany, to provide feedstock for methanol production. It is thought to have excellent prospects in the many European and other countries with large lignite deposits.

In *entrained-bed* gasifiers, no bed exists as pulverised coal particles flow in the gasification agent. Temperatures and throughput are high, with little by-product formation and no problems of coal caking. Operational flexibility is considerable and most types of coal can be used. Most ash is collected as a liquid slag, which can produce heat losses, but considerable quantities of this and unreacted coal can be carried over. Materials handling can also be difficult, whilst consumption of gasifying agents is high.

The *Koppers-Totzek* entrained gasifier is already extensively used, especially where a hydrogen-rich, methane-free synthesis gas is required, as in ammonia production. However, its operation at atmospheric pressure means that present models have a lower thermal efficiency than their Lurgi rivals. A pressurised version is being developed by its owners, Krupp-Koppers.

Future Technologies

Second-generation successors to presently available gasification technologies are not likely to be in commercial use until the late 1980s. All

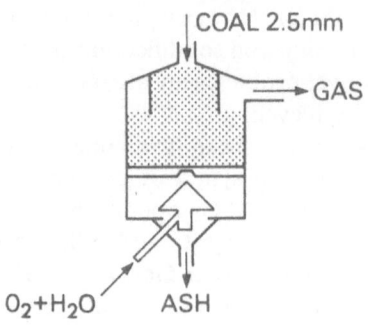

MOVING BED

COAL 2.5mm

→ GAS

O_2+H_2O ASH

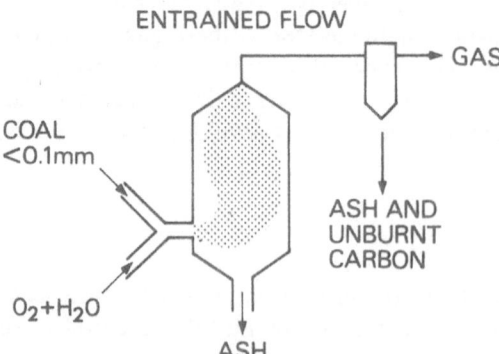

FLUIDISED BED

→ GAS

COAL
0.5mm

ASH AND
UNBURNT
CARBON

O_2+H_2O

RESIDUAL CHAR

ENTRAINED FLOW

→ GAS

COAL
<0.1mm

ASH AND
UNBURNT
CARBON

O_2+H_2O

ASH

FIGURE 2.2 Alternative coal gasification technologies

SOURCE IEA Coal Research, Economic Assessment Service.

share the characteristic of high operating pressures, to exploit the advantages of higher throughput, easier gas handling and the ability to use product gases without further compression, and most are oxygen blown, to maximise calorific value. The ability to use a wide range of coals is also an important design feature, as is operating simplicity. When they are introduced there should be improvements in the thermal efficiencies and costs of coal gas production. The new technologies which are most advanced include:

The *slagging-Lurgi* gasifier, in which modifications to the basic Lurgi design allow liquid ash to be collected, thus removing the limits on operating temperature. This reduces oxygen and steam consumption, increases the amount of carbon monoxide in the final product gas and creates less tar and oil. These by-products are also recycled to the gasifier, lowering the cost of waste treatment. The British Gas Corporation have operated a 350t/d demonstration plant at Westfield, Scotland, since the mid-1970s, and this is being developed to a commercial scale. British Gas have also developed a new High Carbon Monoxide (HCM) process to upgrade the slagging-Lurgi output into SNG.

The *Shell-Koppers* gasifier is a development of the Koppers-Totzek design by Krupp-Koppers and Shell. It operates at higher temperatures and pressures than the original, and reduces the problem of solid carry-over. Its main use will be as a source of standard synthesis gas for refinery feedstock, using any type of coal. A 100t/d pilot plant has been operating at Shell's Harburg, West Germany, refinery since 1978, but a planned 1000t/d demonstration plant has been cancelled. Commercial operation is expected to be in dual modules, each gasifying 2 500t/d of coal.

The *Texaco* entrained gasifier is based on proven technology for the partial oxidisation of heavy oils. Coal enters the gasifier as an aqueous slurry, which reduces feed problems and allows coal fines to be used, although at a cost of greater heat requirements for evaporation of water, and higher carbon dioxide levels in the product gas. Successful pilot plant operation at Texaco's 15t/d Montebello, California, laboratory has led to a 1 000t/d demonstration plant being built by Southern California Edison at Coolwater, California. When completed in 1983, it will be adapted to fire a combined cycle electricity unit. A 200t/d plant is also being built by the Tennessee Valley Authority (TVA) at Muscle Shoals, Alabama, to provide synthesis gas for fertiliser production, and a similar sized unit has been built by Dow Chemicals at its Plaquemine, Louisiana, plant. The process is also likely to be used as a source of gas for direct reduction of iron ore. Other pilot plants are being operated by Ruhrkohle in West Germany and Creusot-Loire in France.

The *U-Gas* process is a single-stage, fluidised-bed gasifier, with recycling of fines and continuous removal of agglomerated ash. Developed by the US Institute of Gas Technology, it is designed to be either air- or oxygen-blown producing a low/medium-CV fuel gas. After successful operation of a Chicago pilot plant, Memphis Light are planning a $700 million 3 100t/d demonstration plant to serve industrial users, although this has been postponed.

The *Hygas* gasifier comprises a succession of fluidised beds, each at a higher temperature than the last. Hydrogen-rich gases produced from the gasification of residual char in the final bed are fed into the first beds, where they directly hydrogenate the coal to form methane (which is also produced by reaction of the remaining hydrogen and carbon monoxide). This process has been designed as a method of making SNG from a variety of coals by the US Institute of Gas Technology, which has successfully operated a Chicago pilot plant, although several problems need to be resolved.

The *Cogas* reactor is unusual in that it produces both gas and liquid products. The initial stages use the COED technology – a series of fluidised beds, which pyrolise and partially gasify the coal – developed by FMC Corporation at a 36t/d Princeton, New Jersey, pilot plant. Further development by the British Coal Utilisation Research Association (BCURA) has added a final stage, in which the bulk of residual char (up to 60 per cent of the feed coal) is gasified with steam in a fluidised bed, which is heated by burning the remaining char in an entrained bed furnace. A utility consortia, with government support, has proposed a 1 200t/d demonstration plant at Perry County, Illinois.

Other technologies whose development is less advanced, but which are thought to show promise for the 1980s and beyond, include:

The *Exxon catalytic* gasifier makes use of a low-temperature fluidised bed. Crushed coal is sprayed with a carbonate (potassium or calcium) catalyst and gasified with steam. The catalyst encourages immediate methanisation of the coal gases, producing a high-methane gas suitable for SNG production, and reduces the caking problems associated with many coals, allowing more types to be used. Although incomplete conversion of the coal occurs, and methane separation has to be carried out by cryogenic distillation, Exxon claims that energy savings are considerable. A 100t/d pilot plant may be operating near Rotterdam in the early 1980s, feeding its gas to an adjacent fertiliser plant.

The *British Gas composite* process couples two technologies together to allow a wide range of coals, and their fines, to be processed. In the first stage, an entrained gasifier reacts pulverised coal (mainly fines), steam and oxygen with the gaseous products, then heating and gasifying the coal in a slagging-Lurgi bed. A small pilot plant is intended for construction at Westfield, Scotland.

Several air-blown gasifiers have also been developed for the specific purpose of producing an LCV gas for combined cycle electricity generation. These include:

The *Allis-Chalmers Kilngas* gasifier has been developed at a 60t/d pilot plant at Oak Creek, Wisconsin. A 600t/d demonstration plant is planned at East Alton, Illinois.

The *Combustion engineering* entrained gasifier burns a third of the coal, using the hot combustion gases to gasify the remaining coal, with the char residues being returned to the combustor. A 120t/d pilot plant has been successfully operated at Windsor, Connecticut, and demonstration plants are planned for West Lake, Illinois and Lake Charles, Louisiana.

The *Ge-Gas* gasifier is a pressurised fixed-bed reactor, which also makes use of coal fines by extruding them into the bed as a tar-bound paste. The US General Electric Co. has successfully operated a 24t/d pilot plant at Schenectedy, New York.

The *Westinghouse* gasifier is a multi-stage, fluidised-bed process, in which char is recycled and dolomite added to fix sulphur. A 15t/d pilot plant has been operated at Waltz Mill, Pennsylvania. Similar LCV processes include the *Babcock and Wilcox, Curtiss-Wright, Foster-Wheeler, Japan Coal Technology Centre* and *NCB* gasifiers.

Others which show promise are the *Battelle-Carbide, Rockwell-Cities Service, Hitachi* and *Saarberg-Otto* gasifiers. By contrast, four favourites of the 1970s – *Bigas, CO_2 Acceptor, Hydrane* and *Synthane* – have all experienced technical and economic problems.

A technology which is expected to be viable in the future is molten bath gasification, in which a molten substance – usually iron or a carbonate – is used to heat the coal, to catalyse reactions and to capture sulphur. These give high gas yields from a wide range of coals, although corrosion problems are substantial. The most advanced processes are the *ATGAS* (Atomics International), *Kellogg, Rummel* (Rheinische Braunkohl) and *SMI* technologies.

Combined Cycle Power Generation

One aspect of coal gasification which has attracted attention is its use in combined cycle electricity generation systems.[20] After purification, low-

or medium-calorific gases are burnt to drive a gas turbine engine to produce electricity and to compress air for the gasifier itself. The temperature of the turbine exhaust gases is then high enough to allow a second, steam turbine, cycle to be operated. The potential coal-to-electricity efficiency of such a combined cycle process is up to 45 per cent, although this is unlikely to be achieved in the first generation of plants. The process also reduces sulphur and other emission levels, and has lower cooling water requirements than conventional power stations, although the high combustion temperatures may result in high rates of nitrous oxide formation. Proposals have also been made to run a combined cycle system on the hot gases emerging from coal-fired pressurised fluidised-bed boilers, and a US demonstration plant is to begin operation in the early 1980s.

Gas turbines have long been used as a flexible means of meeting peak electricity requirements, and some 15 000MWe of (non-coal) combined cycle is already in operation around the world. However, reliability problems – which would be exacerbated by impurities in coal-derived gases – have hindered their use for base-load purposes, although new materials and better methods of cooling turbine blades should change this situation. Present maximum turbine inlet temperatures are 1 100°C, which are sufficient to allow economic operation. Greater efficiencies will be obtained as inlet temperatures approach 1 500°C, as is expected to happen by 2000. A 170MWe plant at Lunen, West Germany, has been successfully operated as a combined cycle by the STEAG utility, and Southern California Edison will add a 110MWe (75MWe gas turbine; 35MWe steam turbine) to the Texaco demonstration gasifier at Cool Water, California, by the mid-1980s. Similar units will probably be fitted to other gasifier demonstration plants.

Fuel Cells

The fuel cell is a method of converting the fuel energy of coal gas into electricity by electrochemical rather than thermal means.[21] It comprises an electrolyte and two electrodes, which are exposed to hydrogen (in low- or medium-calorific gas) and oxygen (in air) respectively. Dissociation of the hydrogen at the anode produces hydrogen ions and electrons, which move towards the oxygen-rich cathode (where they combine to form water) by separate routes – the ions through the electrolyte, and the electrons through an external circuit as direct current. Waste heat from the hot coal gases can also be used to drive a steam turbine. Current fuel-cell technologies use phosphoric acid and these have been adapted for use with coal gasification. Coal-to-electricity efficiencies of 35–40 per cent are expected, although performance is affected by carbon monoxide. A 3.5MWe unit has been built near to New York by Consolidated Edison. In the long term, molten carbonate electrolyte is expected to be used, giving coal-to-electricity efficiencies of up to 50 per cent.

Both these designs should be non-polluting and can operate efficiently at any level of loading. When using medium-calorific gas, they can be sited up to 400km from the gasifier, and it is expected that they will ease the environmental problems of electricity production from coal.

LIQUEFACTION

Hydrocarbons which are liquid at normal temperatures have a high ratio of hydrogen to carbon: the aim of all liquefaction processes is therefore to add hydrogen to, or remove carbon (and other impurities) from, the coal substance.[22] To enable either of, or both, these reactions to occur, the complex organic molecules of the coal must be broken down by thermal and/or catalytic chemical reactions. In the 'synthesis' route, gasification technologies are used to reduce the coal to its simplest units – essentially carbon monoxide and hydrogen – which are then recombined to form the desired products. In 'degradation' routes, the coal molecules are only partially broken up, so that the desired products are reached more directly. The simplest way of achieving this is pyrolysis, although this gives low liquid yields and leaves excess amounts of char. Other, hydroliquefaction, processes hydrogenate the coal with elemental hydrogen and/or a 'donor' solvent, and extract liquid products in an organic solvent. These are then separated by distillation, with the solvent being recycled. A distinction is usually made between solvent extraction and direct liquefaction techniques, according to their use of donor solvents or direct, catalytically-encouraged coal–hydrogen reactions (although the coal ash has a limited catalytic effect in all processes). The greater the amount of hydrogen which is added to the coal, whether directly or via a donor solvent, the higher the yield of liquid products – although at a cost penalty.

Both synthesis and degradation technologies were extensively used before, and during, the Second World War, but only one method – Fischer-Tropsch synthesis – has since remained in large-scale use (see below). Although degradation routes are usually more energy efficient than synthesis (because they only partially break the chemical bonds of the coal molecules) and might be expected to be commercially dominant by 2000, the proven nature and commercial availability of synthesis will favour its use in the short–medium term. However, introduction of either will be hampered by the fact that, for economic and technical reasons, their products differ from those derived from crude oil. For many uses, this will require investment in new refining and chemical processes, or a 'division of markets' between coal and oil liquids.

Synthesis

During the 1920s, Fischer, Tropsch and other German collaborators discovered the temperatures, pressures and catalysts which were necessary to convert coal gases into a mix of hydrocarbons and other organic compounds (see Fig. 2.3). Nine plants using this Fisher-Tropsch technology were built in Nazi Germany, and a successor – SASOL 1 – has been operated since 1955 at Sasolburg, South Africa, by the South African Coal, Oil & Gas Co. (SASOL).

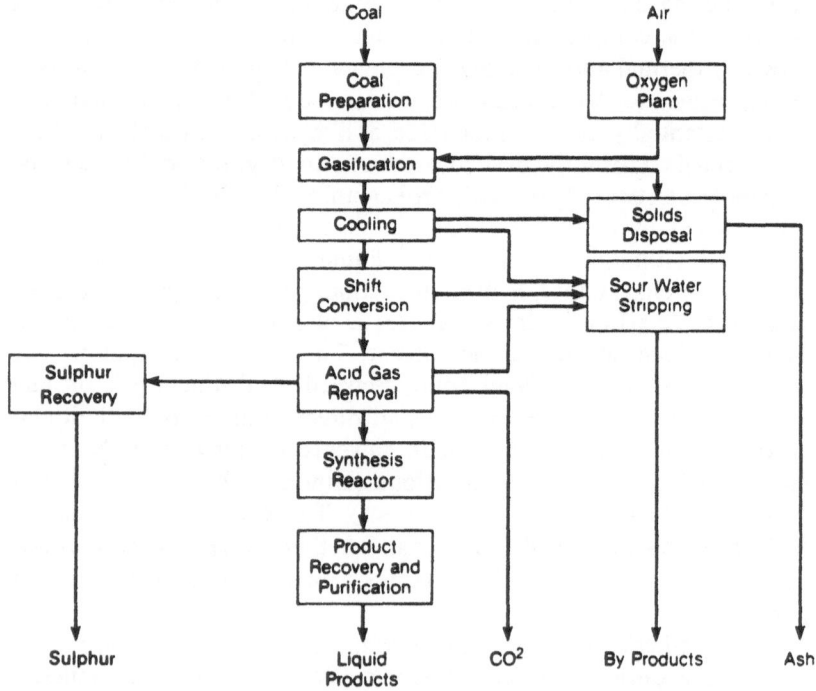

FIGURE 2.3   Coal liquefaction by synthesis: generalised flow plan

SOURCE   US Department of Energy.

The *SASOL* technology uses Lurgi gasifiers to produce synthesis gas. In SASOL 1, which uses 5.5mtpa of coal, two processes – *ARGE* (using 60 per cent of the gas) and *SYNTHOL* (using the rest, plus tail gases from ARGE) – convert it, over modified iron catalysts, into a mixture of hydrocarbons, alcohols, carbonyl compounds and acids. As SYNTHOL creates lighter products, this has been the only process used in the successor plants, SASOL 2 and SASOL 3. These are located at Secunda and, when fully operational in the mid-1980s, will provide half of South Africa's liquid fuel needs from 27mtpa of coal. The scale of these plants is indicated by the components of SASOL 2, which includes: six boilers; six oxygen plants; thirty-six Lurgi gasifiers; four Rectisol gas purification units; seven SYNTHOL units; chemical and oil workup sections; catalytic reforming and polymerisation units; an isomeration unit; a primary fractionater; a vacuum distillation tower and various hydrotreaters. Further units will be added during the 1980s. Several plants using SASOL technology, of which Fluor owns international rights, may be operating in Australia and the USA by 1990. However, their large-scale introduction will be hampered by a high capital cost, a lack of flexibility in product range and the costs of waste disposal.

The *Mobil-M* process is a semi-proven synthesis route which is considered to have great promise. Synthesis gas from coal is first converted to methanol, using one of several existing technologies such as the ICI low-temperature route. Although methanol can be used directly as an 'extender' to motor fuels, or as a chemical feedstock, this process converts it, over a zeolite (crystalline aluminosilicate) catalyst, into high-octane gasoline liquids, with yields of up to 88 per cent. After successful operation of a 4b/d methanol-gasoline plant in New Jersey, a 100b/d demonstration plant is being built near to Cologne by Mobil and Rheinische Braunkohlenwerke, with US and West German government support. A 13 000b/d plant, using natural gas feedstock, is also being built in New Zealand.

Degradation

Most degradation technologies (see Fig 2.4) are based on the work of Bergius, whose studies of coal hydrogenation led to the construction of commercial plants in Germany, which operated twelve during the Second World War, and the UK. In the USA, four developments of these earlier processes have reached commercial demonstration status:

The *SRC-1* process mixes pulverised coal with an anthracene oil solvent and hydrogenates the resulting slurry. After critical solvent

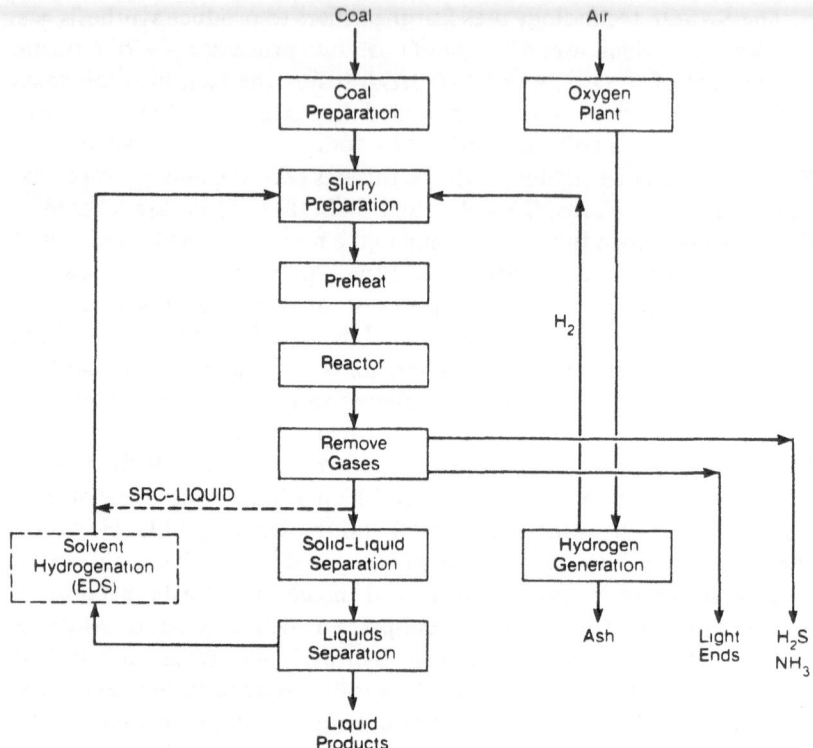

FIGURE 2.4   Coal liquefaction by degradation: generalised flow plan

SOURCE   US Department of Energy.

deashing (introduced after difficulties with filtration) and distillation of the solvent for recycling, a brittle, pitch-like solid, with a low sulphur content, is obtained. A 6t/d plant at Wilsonville, Alabama, was to have been followed by a 6 000t/d plant at Newman, Kentucky, producing the equivalent of 20 000b/d of oil, but this has been delayed by US government cutbacks.

The *SRC-2* process is a development of SRC-1, which uses more severe hydrogenation conditions to create higher yields of liquids. Typically, 70 per cent of output is in the form of fuel oil and naphtha, with the rest as gaseous hydrocarbons. The separation problems of SRC-1 are avoided by distillation. A 50t/d plant at Fort Lewis, Washington, was successfully modified to SRC-2 operation by a Gulf Oil subsidiary, and should have been followed by a 6 000t/d demonstration plant at Morgantown, West Virginia, financed by the US, West

German and Japanese governments. These plans have been delayed due to US government cutbacks. Research has also been carried out in Japan, and Japanese-led consortia plans to operate a commercial plant in Australia, using Victoria lignite deposits, by the late 1980s.

The *Exxon donor solvent* process uses a 'donor' solvent, which has been separately hydrogenated to avoid catalyst poisoning. Typical products are a heavy fuel oil, suitable for fluid coking, a heavy naphtha fraction and light hydrocarbon gases, which are converted to hydrogen. Naphtha output can be maximised, but at the cost of reducing thermal efficiency. Exxon successfully operated a 1t/d pilot plant at Baytown, Texas, which has been followed by a 250t/d demonstration plant on the same site. Other participants in the project include the Phillips, Arco and Agip (Italy) oil companies, the Electric Power Research Institute, Japan Coal Liquefaction Development Co. and Ruhrkohle, with financial support from their respective governments.

The *H-Coal* process, originally developed as a means of removing sulphur from fuel oil, directly reacts coal and hydrogen in the presence of a disposable cobalt–molybdenum catalyst. This produces either fuel oil or a low-sulphur 'syncrude', plus hydrocarbon gases and a carboniferous residue. Hydrocarbon Research has operated a 3t/d pilot plant at Trenton, New Jersey, and has planned a demonstration plant at Cattletsburg, Kentucky. This would use 250t/d of coal in its syncrude mode, and 600t/d for fuel oil. Other participants include the Amoco, Ashland, Conoco and Mobil oil companies, the Electric Power Research Institute and Ruhrkohle, with further support from both national and state governments. Other, less advanced, US technologies include:

Conoco   An advanced hydrogenation process, which feeds a coal-solvent slurry into a molten zinc chloride bath, where hydrogenation occurs. This provides a rapid reaction and gives high yields of gasoline-type liquids.

Lummus   A two-stage process which separates the primary coal liquefaction stage from the secondary catalytic hydrogenation, giving more selective hydrogen use and higher and more flexible liquid yields.

Occidental   A hydropyrolysis process, in which coal is rapidly pyrolised in an entrained bed and inert atmosphere, producing gaseous and liquid fuels. The quality of liquid products is improved by quenching the hot product vapours in a hydrogen-rich liquid solvent. Work is proceeding at an Irvine, California site.

US Steel clean coke   In this process, part of the coal charge is carbonised, and the other slurried and hydrogenated. The first stage produces metallurgical coke and intermediate products, which are fed into the hydrogenation reactor. A demonstration plant is planned, using SRC-2 technology for the hydrogenation stage.

Other US processes which have reached pilot-plant status, but encountered severe problems, are *Consol* and *Synthoil*.

Extensive work on coal liquefaction is also taking place in West Germany, based on its previous experience in this area, and some observers believe that it will be the first OECD country to establish a commercial liquefaction industry.[23] The main processes are:

Ruhrkohle-Veba A direct successor to Bergius technology, this method hydrogenates a coal-solvent slurry, and separates the liquid products by distillation. Residual matter is then gasified in a Texaco gasifier. A 200t/d pilot plant is situated at Bottrop, West Germany, and Ruhrkohle plans to operate a 6mtpa plant, providing 10 per cent of West German liquid fuel needs, by 1990.

Saarberg A similar technology to that used by Ruhrkohle-Veba, this is being tested by Saarbergenwerke at a 6t/d pilot plant at Fürstenhausen, West Germany.

In the UK, the National Coal Board has also developed two processes:

The *supercritical gas extraction* method uses a light organic solvent, such as toluene, in the form of a supercritical gas. This increases its solvent properties and causes coal liquids to vapourise more easily. These liquids are released when the solvent is depressurised outside the reactor. Although liquid yields are relatively low, they are of high quality, and the process does not require expensive hydrogenation or separation. After successful operation of a small pilot plant, a 25t/d unit is planned for Point of Ayr, Wales, in the 1980s.

The *solvent extraction* method hydrogenates a coal-solvent slurry. Liquid products are separated and further hydrogenated to give a light feedstock for further refining. A 25t/d pilot plant is planned at the Point of Ayr site during the 1980s.

Extensive liquefaction research has also been undertaken in Japan:

The *Cherry* process catalytically hydrogenates a pulverised coal-heavy oil mix, producing a range of distillates and a carboniferous residue. A 40t/d pilot plant is being operated by the Osaka Gas Co.

The *solvolysis* process dissolves coal in asphalt, producing a pitch-like material for direct use or upgrading. The Japan Electric Power Development Co. has a 40t/d pilot plant at Nagasaki.

Mitsubishi, Mitsui, SMI and Sumitomo have also undertaken research on their own and US technologies.

In the USSR, liquefaction work has concentrated on the joint processing of coal and highly aromatic oils, with the latter acting as a donor solvent. One process rapidly pyrolises a coal–oil mix. Another, designed for use with lignite, hydrogenates a coal–oil paste in the presence of an iron–molybdenum catalyst, producing boiler fuels and distillates with a relatively low hydrogen consumption. Poland is also developing both pyrolysis and direct hydrogenation technologies.

Although much progress has been made with degradation technologies, many problems need to be resolved before commercial use is possible. Major technical difficulties which require further research are the handling and separation of three-phase mixtures of solids, liquids and gases; formation and handling of coal slurries; processing of coal liquids, including hydrocracking, and their characterisation in the same terms as petroleum products. Favourable economics will also require more optimal use of hydrogen. Finally, the surplus heat energy which is produced at most degradation and synthesis plants suggests that liquid yields can be further improved.

COAL CHEMICALS

The by-products of coal carbonisation, together with synthesis gas produced from coal gasification, once formed the basic raw material for organic chemical production.[24] Although some use is still made of coal feedstocks for this purpose, their use is now negligible by comparison with oil and natural gas.

Carbonisation creates many products, which are usually grouped into coke, gases, tars and oils, crude benzole and ammonia liquor. Their yield is determined by temperature, rate of heating, type of oven and the class of coal used.

Coke can be reacted with lime to produce calcium carbide, which can be used to produce acetylene. This is an important intermediate for plastics and other products, but is now only rarely derived from coal, although this may change in the future.

Coal gases, often maximised by the use of gasifying agents such as steam, have been used as a source of hydrogen or carbon monoxide. The latter is converted to valuable intermediates, glycols and metal carbonyls, which play an important role in mineral purification. Coal gases can also be directly synthesised, with ammonia and methanol as the most common products. Methanol is further processed to produce formaldehyde, methyl esters, amines, solvents and microbially-produced proteins. Synthesis gas can also be reacted with olefins, using OXO catalytic technology, to produce alcohols, aldehydes and other intermediates.

Coal tars are fractionated to produce pitch, creosote and a range of oils. Pitch is a valuable source of carbon binders and fillers and electrode coke, whilst creosote and its derivatives are used for timber preservation and plant protection. The oils are highly aromatic and still provide much of the world's production of naphthalene, anthracene and similar products. These are intermediates in the manufacture of a wide range of dyestuffs, resins, disinfectants, pharmaceuticals and plant-protection agents. The lighter tar fractions are refined in the same way as crude benzole, to produce naphtha and light aromatics such as toluene, xylene and benzene. The latter is an important intermediate for cyclohexane, a raw material for nylon production.

COAL MARKETS

In the past, coal was the main fuel in most energy markets and a basic raw material in the chemical, gas and metallurgical industries. Today, its use is confined to one dominant – electricity generation – and two minor – industrial heat and metallurgical – purposes.[25] In the two former markets, coal is in direct competition with other fuels and requires a price differential to offset its higher operating and transport costs. As a result, this 'steam coal' is consumed within regional markets, based on indigenous reserves which are widely distributed on the earth. In metallurgical use, coal can only be partially substituted and this, together with a concentration of reserves, has produced continental or global markets, and higher prices for the coal itself.

Electricity Generation

Coal use for electricity generation was the only rapidly-growing market for coal in the 1960s and 1970s, when it increased at a rate of 4.2 per cent per annum, although this was less than the rate of growth of electricity demand. By 1979, this market accounted for approximately 65 per cent of coal use, mainly in North America, Western Europe, the USSR and China. In the past, coal's main competitors have been oil and, to a lesser extent, gas. As these are less costly to handle and use, delivered coal prices needed to be 4–15 per cent below that of oil, and 5–22 per cent below that of gas, to be economic in 1979. In fact, the oil and gas price rises of the 1970s made these sources uncompetitive in many countries, and nuclear power stations have become the main rival to coal for base-load electricity-generating capacity. The economics of these two routes have been subject to great controversy but, in general, utilities and governments have concluded that nuclear electricity is the most economic.[26]

The disputes rest on the assumptions which are made. One crucial factor is the discount rate used, with low rates favouring nuclear power and high rates coal. Another is predicted increase in capital costs over the construction period. One survey of European and American estimates found that the figures for coal were broadly comparable, but that those for nuclear power varied by 45 per cent – and by up to 100 per cent on other, private, costings.[27] Anticipated load factors are also important, as high usage allows the capital cost to be spread over a greater output. Early estimates of nuclear reactors used a load factor of 75–85 per cent, which have proved unrealistic. In practice, load factors for nuclear capacity have usually been less than the 60–65 per cent achieved for coal, despite the reduction due to use of older coal-fired capacity for peak rather than base-load generation, an option which is not available for nuclear plant. Finally, it appears that nuclear power has received a greater amount of public support for research and development, insurance guarantees, waste disposal and so on, than has coal. One US study found that, to 1976, nuclear had received $15.3–17.1 billion, and coal $6.8 billion in Federal government incentives.[28]

An authoritative study of coal–nuclear competition has found that the break-even price for coal is 1.6 $/GJ with sulphur control, and 2.4 $/GJ without sulphur control.[29] (see Fig 2.5) This, together with political restrictions on the use of nuclear power, will ensure that coal usage in this market continues to increase. An Exxon study forecasts that utility coal demand in the West will rise by 3.7 per cent per annum until 1990, and 2.9 per cent from then until 2000.[30] The WOCOL Report projects that coal use in OECD power stations will rise from 600 mtce per annum in 1977 to 1 325–1 850 mtce per annum by 2000.[31] In the short-medium term, coal use may also be boosted by conversion

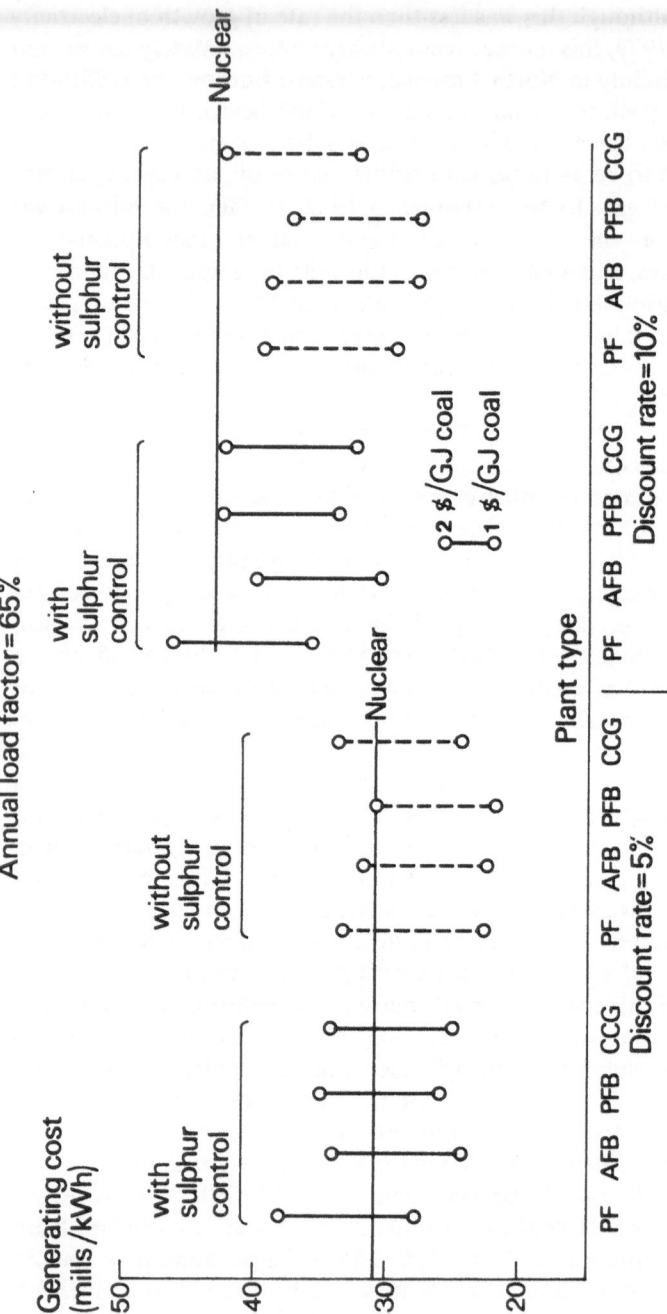

For key see Table 2.1

FIGURE 2.5   Costs of electricity from new coal technologies and nuclear power

SOURCE   IEA Coal Research, Economic Assessment Service, 'The Economics of Coal-Based Electricity Generation', (1979).

of oil- or gas-fired plant to coal and use of coal-oil/methanol mixtures instead of fuel oil.

However, the realisation of these targets will depend on the performance of the world economy and, in particular, whether Western economies move out of the post-1979 recession and are able to grow in the 1980s without suffering further 'oil shocks'. High prices may further reduce growth in energy and electricity demand. Coal prospects will also be determined by the movement of coal prices – which may rise in real terms as producers increase their economic rent – and the introduction of new technologies, particularly fluidised bed combustion and combined cycle electricity generation. These are said to offer economic advantages, especially where sulphur removal is required, but forecasts depend upon small changes in processes parameters, such as thermal efficiency, capital and waste disposal costs and construction time. In general, an atmospheric FBC power plant is expected to show a 4 per cent improvement in thermal efficiency over, and an 11 per cent lower coal requirement than, a conventional pulverised-fuel plant with a wet-lime scrubber. Combined cycle with either a pressurised FBC, or gasification, is expected to produce improvements of 6–11 per cent in thermal efficiency, and 16–25 per cent savings in coal requirements. Atmospheric FBC is likely to be economic at low or moderate coal prices, and combined cycle at higher with the pressurised FBC version likely to be preferred where sulphur removal is not required[32] (see Tables 2.1 and 2.2).

Metallurgical

About 20 per cent of world coal output is used for metallurgical purposes, largely in iron and steel manufacture. Until recently, this has largely been carried out in open-hearth and basic oxygen furnaces where coal – in the form of coke – is required, with no substitution being possible, except to a limited extent for the provision of heat. The use of oil or gas for this latter purpose kept the annual increase in metallurgical coal use to 0.6 per cent per annum in the 1960s and 1970s, but the rate of growth is forecast to rise at 2.5–3 per cent per annum until 2000 by Exxon. The WOCOL Report projects a more modest growth in OECD countries, but this will be offset by expansion in Third World steel industries. However, these figures may be reduced by the introduction of new steel-making technologies, particularly direct reduction and electric-arc methods, although these will partially substitute the combustion or gasification of steam coals for the coking of metallurgical coals.

TABLE 2.1   Capital cost and thermal efficiency of coal–electricity technologies

System	High S coal		Low S coal	
	Cost[a] ($/kWe)	Efficiency (%)	Cost ($/kWe)	Efficiency (%)
Pulverised fuel (PF)[b]	650	33.0	545	36.5
Atmospheric fluidised-bed (AFB)	505	35.8	482	36.5
Pressurised fluidised-bed (PFB)	586	39.2	529	40.2
Gasification/combined-cycle (CCG)				
Lurgi	702	35.0		
General Electric (GE)	630	39.0		
Combustion Eng. (CE)	708	38.1	Not estimated	
Texaco (air)	625	38.1		
Westinghouse (W'hse)	649	41.8		

[a] Capital costs were derived from published North American sources by comparing costs for major plant items. They include a 10 per cent architect/engineer's fee, and a project contingency allowance of 15 per cent. They were compared with European estimates for pulverised fuel generation to ensure applicability of the results. Prices are mid-1978.
[b]   High S case assumes 85 per cent sulphur removal. Low S case assumes no sulphur removal.

SOURCE   IEA Coal Research Economic Assessment Service, 'The Economics of Coal-Based Electricity Generation', (1979).

Industrial

Industrial use of coal has fallen in recent decades, but the WOCOL Report projects a growth rate of 5–7 per cent per annum until 2000. Exxon forecasts a similar rate to 1990, rising to 9.7 per cent per annum between then and 2000, although this figure includes synfuel production. At present, the main end-uses of coal in this sector, excluding the specialised needs of iron and steel production, are process heat, direct heat below 600° C and space heating, all of which can be provided from a single boiler.[33] Other requirements are not generally amenable to direct coal use, with the exception of high-temperature direct heat for the firing of brick and cement kilns.

As the capital and operating costs of coal-fired boilers are higher than for other fuels, considerable coal-oil/gas price differentials are required. These have been estimated at 20–30 per cent for the installation of new boilers, and 70–100 per cent for the scrapping of existing, adequate equipment, and replacement with coal.[34] One important factor which determines coal's share of this market is the size of boiler installation, with coal usually being more competitive for larger capacities, where pulverised-fuel equipment can be installed. The introduction of FBC boilers is expected to increase the attractiveness of coal as an industrial fuel. Although coal-derived electricity and gas may substitute for some direct use in the future, combustion in boilers is expected to remain the dominant industrial use until the next century.

Conversion

Conversion of coal to gaseous, chemical and, to a lesser extent, liquid products was once a considerable market. Many observers believe that coal will again be used for these purposes in the future, as supplies of oil and gas are reserved for premium markets. Exxon has predicted that OECD conversion of coal will require approximately 500mtce per annum by 2000, from which 4mbdoe of liquid products will be produced. By the early twenty-first century, the company sees US production alone reaching 7mbdoe, requiring an investment of up to $400 billion (at 1980 prices).[35] The WOCOL Report has projected an output of 0.6–2.8mbdoe by 2000, creating a demand for 75–335mtcepa of coal, whilst the US synfuels programme has aimed to reach a target of 1mbdoe of coal liquids by the 1990s. However, other forecasts are more pessimistic, arguing that the first generation of plants will be plagued by equipment and operating problems, with correspondingly low load factors.[36] As capital costs account for the largest element in product prices, this may delay economic production of liquids and gas until the 1990s or afterwards. Formidable problems are also foreseen in supply of raw materials, especially oxygen, equipment manufacturing capacity, availability of skilled labour and environmental acceptability. Output will also tend to be concentrated in a few areas, according to the mining costs and characteristics of coal deposits. In general, gasification and synthesis technologies will favour the use of lignite and sub-bituminous coals, while direct liquefaction technologies will work best with coals of high hydrogen and volatile matter content.

In coal gasification, process costs are typically two–three times that of the coal itself.[37] As a result, the economic ranking of different gasifier technologies are not affected by changes in coal price. However, environmental costs are expected to be substantially more onerous for low-calorific gasifiers, and will be an important factor in determining their viability. Despite this, one study has predicted a substantial market for low-calorific gasifiers serving adjacent industrial estates.[38] Medium-calorific gas may be produced initially for specialist uses, as in Western

TABLE 2.2 Summary of coal input and operating costs for coal–electricity technologies ($/kWe)[a]

Item	High sulphur coal								Low sulphur coal		
	PF[b,c]	AFB[d]	PFB[c]	Lurgi	GE	CE	Texaco	W'hse[c]	PF	AFB	PFB
Coal cost $ S1/GJ	10.91	10.06	9.18	10.29	9.09	9.45	9.45	8.61	9.86	9.86	8.96
Coal cost $ S2/GJ	21.82	20.11	18.37	20.57	18.18	18.90	18.90	17.22	19.73	19.73	17.91
Coal cost $ S3/GJ	32.73	30.17	27.55	30.86	27.27	28.35	28.35	25.84	29.59	29.59	26.87
Sorbent cost	0.56	1.21	1.32	–	–	–	–	1.24	–	–	–
Waste disposal cost	1.12	0.77	0.64	0.19	0.17	0.18	0.18	0.60	0.18	0.18	0.17
Operating labour cost	0.63	0.49	0.53	0.71	0.65	0.53	0.55	0.61	0.45	0.45	0.49
Maintenance cost	2.28	1.77	2.06	2.47	2.21	2.49	2.20	2.28	1.91	1.69	1.86
Insurance etc. cost	3.42	2.66	3.09	3.70	3.32	3.73	3.29	3.42	2.87	2.54	2.79
Total non-fuel costs	8.0	6.9	7.6	7.1	6.4	6.9	6.2	8.2	5.4	4.9	5.3

[a] All costs relate to a 1 000MWe plant operating at a load factor of 0.65 (that is, 5 694GWh/year), at base investment level. Coal feed: ash, 10 wt per cent; sulphur, 3.5 wt per cent for high sulphur coal; higher heating value, 26.75GJ/t (as received).
[b] For key, see Table 2.1.
[c] Ca/S mole ratio = 1.5.
[d] Ca/S mole ratio = 3.5.

SOURCE IEA Coal Research, Economic Assessment Service, 'The Economics of Coal-Based Electricity Generation' (1979).

Europe where it will probably be used to dilute very high calorific-value natural gas from offshore, or overseas, deposits. In both cases, production costs should fall substantially when economies of scale are possible.

The commercial viability of the only large-scale liquefaction plant in the world, SASOL 1, and of its two successors – SASOL 2 and 3 – are controversial. Although its operator claims that it has been profitable in the late 1970s, it is also true that the plants have been built mainly for strategic reasons and that their output is partially subsidised by the South African government. However, several reports have suggested that, in the short–medium term, synthesis routes are likely to be more economic than degradation, although in the case of the Mobil-M process, the methanol feedstock may be more cheaply obtained from natural gas than coal. Although individual technologies will be favoured for the particular product mix and quantities they produce, in general synthesis routes are likely to be favoured for highly refined products, while degradation routes will be used to provide substitutes for fuel oil.

In both cases, capital costs are likely to be over half of the product price, making their viability heavily dependent on construction and financing arrangements. The costs of hydrogen production will be another important factor, with initial evidence suggesting that gasification of the carboniferous residues from the plants will not be the most economic method. As with gasifiers, environmental controls may also be costly, perhaps amounting to as much as 25–29 per cent of the product price.

Chemical production from coal has declined in importance since the Second World War and now accounts for only a small proportion of organic chemical output. In the 1970s, some 12mt per annum of coal tar was refined, mainly in the steel-producing areas of Western Europe, USSR, USA and Japan. Although the continued availability of coal tar as a cheap by-product of coke production will ensure the industry's survival on its present scale, no major expansion is likely. Greater promise lies with the synthesis of coal gases into desired chemicals, which is already carried out on a large scale in South Africa and – for ammonia and fertiliser production – in some COMECON and Third World countries. The pressure of population on food resources, and the rising prices of gas or oil feedstocks, is likely to substantially increase this latter use, especially in Third World countries with coal reserves. In the longer term, refining of the syncrudes and distillates produced by coal degradation may also become important, but these will suffer from the same economic constraints as the liquid products until the next century. Finally, a considerable proportion of future sulphur output is likely to be derived from coal desulphurisation equipment.

In practice, the economics of many conversion processes are likely to be improved by their being linked together – in the same way as the present chemical industry has clustered around oil refineries – in major 'coalplexes'.

3 Coal, Environment and Health

Concern at the environmental effects of coal production and use is not new. As long ago as 1257, the English Queen Eleanor is reported to have left Nottingham as a result of smoke from coal-burning and, in 1306, the practice was banned on pain of death by Royal proclamation – apparently without success. Control measures were introduced in many European countries during the centuries which followed, and conditions were sometimes attached to mining leases – as with an English colliery of 1791, whose responsibility for reclamation was specified in considerable detail, such as an instruction that the land be 'sown with Rye grass seeds'.[1]

Notoriously, such environmental considerations were thrown to the winds in the industrial expansion of the nineteenth century, when coal production expanded by many orders of magnitude. The effects of this precipitate growth – scarred landscapes overshadowed by brooding mountains of waste, grimy cities with countless belching chimneys – formed an image of coal which colours opinion even today. In fact, much progress has been made in dealing with its worst effects, although increased awareness and knowledge of environmental costs means that the problem will be a major constraint on coal production and use in the future.[2]

COAL MINING

Like most industrial activities, coal mining produces many different impacts on the surrounding environment, with the importance of these depending on such factors as the type of mining practised, the geography and ecology of the area affected and public attitudes towards them[3] (see Fig. 3.1). The most controversial aspect of modern coal mining, however, is one which is unique to extractive industry – the need to deal with enormous volumes of waste material which is mixed with

The Future of Coal

the coal (see Table 3.1). In both underground and surface mining, these are normally processed in close proximity to the extraction zone – to permanent storage in tips or reservoirs in the former, and to temporary storage, with subsequent replacement, in the latter. As the characteristics of both spoil and surface can vary greatly, even within a single mine, a variety of techniques have been developed to solve the problem. A general distinction can be made between those of restoration, in which land is returned to its original topography and use, and reclamation, which creates different uses and topographies.

Underground Mine Waste

Underground mines produce several types of waste, including that from drivage of shafts and tunnels to reach the coal measures, material from within the seams, material from the roof and floor which is brought to

TABLE 3.1   Land affected by coal utilisation

Facility	Acres over 30-year period
Mining: 10^{15} Btu per year[a]	
Western strip	15 000–96 000
Central strip	216 000
Central room and pillar[b]	525 000
Eastern contour	470 000
Eastern room and pillar[b]	560 000
Waste disposal[c]	15 000
Combustion/Conversion:[c] Input of 10^{15} Btu/year	
Power plants	13 600
Lurgi gasification	6 500
Synthoil liquefaction	6 300

[a] For a high heat value coal (for example, Eastern coal with 12 000 Btu/lb) this is equivalent to approximately 42 million tons per year. For a low heat value coal (for example, Western coal with 8 500 Btu/lb) this is equivalent to approximately 59 million tons per year. For a medium heat value central coal of 10 000 Btu/lb, this is equivalent to 50 million tons per year.

[b] Includes undermined land which is potentially subject to subsidence.

[c] A 1 000 MWe (megawatt-electric) power plant would 'consume' approximately 59 × 10^{12} Btu/year; thus it would take nearly 17 such plants to consume 10^{15} Btu/year. Similarly to use 10^{15} Btu of coal per year would require about 8 Lurgi gasification plants of 250 million cubic feet per day and about 3 Synthoil liquefaction plants each producing 100 000 barrels per day.

SOURCE   Congress of the United States, Office of Technology Assessment, 'Direct Use of Coal' (1979).

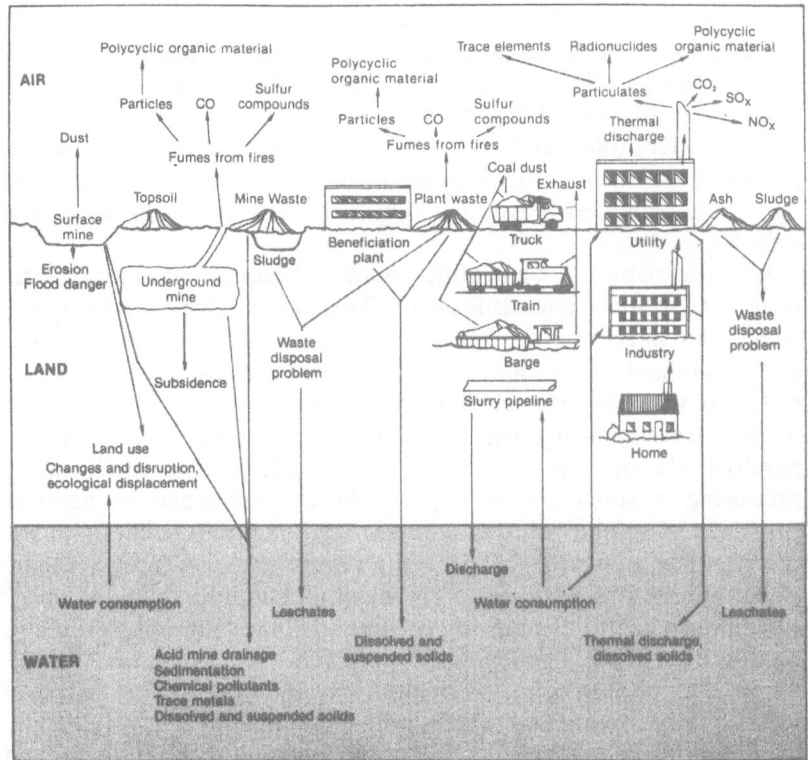

FIGURE 3.1 Environmental disturbances from coal-related activities

SOURCE Congress of the United States, Office of Technology Assessment, 'Direct use of Coal' (1979).

the surface in 'run of mine' coal and residues from coal preparation plants.[4] The percentage of mineral matter in 'run of mine' coal has roughly doubled since the 1950s, primarily due to the less selective cutting of mechanised equipment. The move towards coal preparation also creates larger volumes of waste, the handling of which presents great difficulties due to the presence of water – up to 95 per cent in the case of slurries containing fine coal particles. This must be removed either by evaporation from lagoons – the waste then being periodically excavated for tipping – or by expensive mechanical means such as filter pressing or cone thickening. As the capacity of lagoons is limited for safety reasons, mechanical methods and chemical thickening – such as the addition of cement – will become more important in the future.

The problem of dealing with present production of waste is compounded by the need to deal with old tips, many of which are in poor condition. In the UK these contain an estimated 2.5 bt of waste, covering 11 000 hectares, while in the Appalachian region of the USA estimates suggest that some 3–5 000 waste banks contain up to 3 bt.[5] In addition to their visual impact, many of these can become structurally unstable after prolonged wetting, especially if slopes are steep and a high proportion of the material is in the form of fine particles. When these conditions coincided at Aberfan, South Wales in 1966, 144 people lost their lives.

Exposure to moisture and atmospheric gases can also lead to chemical instability within the waste material. Toxic materials may be released, while the oxidisation of other components, notably pyrites, produces acid compounds. These may be leached into local water supplies. Such reactions produce heat, leading to combustion of the coal particles which have remained within the material. It is estimated that several hundred US tips are undergoing such 'spontaneous combustion', producing a small but locally significant percentage of national emissions of sulphur dioxide (1.8 per cent), fine particulates (1.4 per cent), carbon monoxide (1.2 per cent), nitrous oxides (1.0 per cent) and hydrocarbons (0.6 per cent).[6] The oxygen-deficient conditions under which much of this combustion occurs can also produce high levels of hydrogen sulphide, ammonia and polycyclic organic matter (POM). Under the pressure of public opinion and legislative action, effective solutions have been found to these problems, although they may require long periods of time to be fully successful. Careful tipping and compaction prevents spontaneous combustion, or can cure it (after re-excavation) when already under way, whilst grading into flatter shapes reduces instability and erosion.[7] This also helps revegetation to reduce the visual impact, although to be fully effective this requires the creation of suitable soil conditions and maintenance for many years. In many parts of the world such graded tips have been put to productive use, usually for leisure pursuits but sometimes as agricultural land.

Several alternatives to on-site tipping exist. One is remote disposal which, although more expensive and creating some pollution problems from the transport of waste, can produce environmental benefits by filling excavations in the receiving area. Usually, however, competition from other sources, such as urban refuse, is too intense to make this option feasible. Backfilling, whereby waste is returned underground and packed into exhausted workings, is used in several mining countries, although on a lesser scale than in the past because of its effects on production and fears that it may damage health by creating extra dust. Improved techniques may become available in the future. A proportion of waste – 10 per cent in the UK and more in some other countries – can also be used as fill material for highway and building construction. With shortages, and increased opposition to the quarrying of traditional sources of fill such as sand and gravel, opportunities for using waste should increase in future, although this may be balanced by a decline in major civil engineering projects. Finally, attempts will be made to reduce the amount of waste produced in coal mining. The two most promising attempts appear to be the use of fluidised-bed boilers to burn the coal which remains in waste, reducing the volume and producing useful heat, and the development of more sophisticated coal-cutting equipment to follow coal seams more accurately.

Surface Mine Waste

In surface mining, the ratio of overburden removed to coal produced can be up to 25:1, but these large volumes of waste can usually be returned to the excavation when mining is completed.[8] However, similar problems to tips are usually experienced, including disruption of groundwater flows, leaching of toxic and other pollutants, and erosion and collapse of unstable slopes. Restoration of the land – which is now a legal requirement in most Western nations – can also be hindered by the destruction of topsoils and subsoils by mining operations.

The best restoration practice is continuous in nature, with overburden from current excavation zones placed in previously cut areas and then, after regrading and some compaction, covered with topsoil and revegetated. One method accepts the resulting soil conditions, chooses plants which are suited to them and allows a natural 'climax' community to evolve. This can be slow, and in many areas the conditions may be so harsh that a few species can prosper. More usually, soil conditions are altered, making it suitable for favoured varieties and giving some choice in after use. Although quicker, this method often requires maintenance for many years, and is therefore more expensive. Factors which affect the success of both methods include level of acidity or alkalinity; availability of moisture; flattening of soil (which can be over-compacted by heavy machinery); temperature; level of nutrients and presence of necessary soil micro-organisms.

In many cases, land can be reclaimed to its original state, and notable successes have been achieved in several European countries as in the lignite fields of West Germany.[9] Where this is not possible, improvements can be made to the land's original condition – such as the creation of level surfaces in undulating country, or hills in flat landscapes. Soil nutrition levels may also be enhanced and stoniness reduced, whilst excessive acidity or alkalinity can be modified. Such improvements usually require some degree of legislative compulsion to become widespread, and may be impossible in certain regions – particularly arid zones such as the Western USA. Often, surface mining in these conditions will result in what has been termed 'national sacrifice zones' and may well be banned in the future.

Water Quality

Less extreme, but still significant, impacts occur in other areas of the environment. Underground mining uses large quantities of water for dust suppression, cooling of equipment and coal washing, and similar tasks are carried out in surface mines[10] (see Table 3.2). The introduction of hydraulic mining, already practised in the USSR, and slurry pipelines on a large scale will further add to this demand. Most mines already treat waste water from these activities, but re-use will probably become normal, both to minimise pollution problems and to reduce water needs, especially in arid areas.[11]

Water quality can also be affected by seepage from abandoned workings, or mined waste. Acid Mine Drainage (AMD) is a particular problem in disused workings, where exposed sulphide minerals react with air and water to form sulphuric acid, which is carried away in groundwater flows. Aquatic ecosystems can be damaged by the increased acidity, and may also suffer from the increased solubility of toxic metals, and other harmful substances, resulting from this. The problem can be minimised by reducing water flows (avoiding undue fracturing of strata and sealing of holes and fissures) and exclusion of air by flooding, although care must be taken that this does not increase rates of seepage. Leaching from mine wastes can be reduced by careful design, such as an impermeable base, and by controlled tipping and compaction. As a last resort, polluted water can be treated by alkali neutralisation and flocculating agents to settle out suspended solids, although this is expensive, only partially effective and produces a hazardous sludge which itself causes disposal problems.[12]

Dust

Further environmental pollution results from the disintegration of mined solids, the particulates from which can produce haze, con-

TABLE 3.2 Water requirements for coal system activities

| Coal activity | Size | Water consumed by energy facility | | Peak water requirements associated with population increases[a] | |
		Acre-ft per year	Gallons/10^6 Btu in product	Construction (acre-ft/yr)	Operation (acre-ft/yr)
Surface mine	12MMtpy	3 400	4	71	323
Underground mine	—	0	0	275	1 490
Power generation	3 000MWe	29 000	54 (thermal) 157 (electric)	853	260
Lurgi gasification	250MMscfd	6 700	28	1 570	350
Synthoil liquefaction	100 000Bbl/day	17 500	28	1 750	1 800
Slurry pipeline	25 MMmtpy	18 400	14	Not considered	Not considered

[a] Assumes 150 gallons per capita per day, a multiplier of 2 to account for added service personnel during construction, and a multiplier of 3.5 to account for families and service personnel during operation.

SOURCE Congress of the United States, Office of Technology Assessment, 'Direct Use of Coal' (1979).

taminate soil and water, discolour buildings and pass through the human nasal passage to enter the lung. Such effects depend upon the size of the particles, the wind speed and direction and humidity, and are at their worst in arid regions where the dust is not washed down by frequent rainfall. In practice, the fine particles which are most damaging to human health settle out quickly and are more a problem to mine employees than the public. Generally, air quality is most affected by surface mines, which expose a large surface area of unstable solids to the atmosphere – the East Gillette mine in Wyoming, for example, emits 819 tpa of particulates.[13] Poor tipping of waste and the air drying of washed coal can also be significant sources of dust. Many countries now have strict controls on particulate emissions from mines, and good operating practice – prompt covering, screening from winds and capturing dust by enclosure or water spraying – can reduce these to acceptable levels.

Noise and Visual Intrusion

Adequate remedial measures are also available for the visual and noise problems associated with coal mines, although these are not always used. Plant, small excavations and tips can be concealed by vegetation screens or grassy banks, an approach which has been particularly successful in the West German lignite mines. In surface mining, impacts can also be minimised by excavation of only part of the land at any one time, and backfilling and restoring this as soon as possible. Where some degree of visual intrusion is inevitable, careful landscaping and design can minimise the effects, especially when such factors as future expansion, ecology, hydrology, topography and human geography are taken into account. Several of the measures which can be taken to reduce visual impact, such as erection of barriers, also reduce the problem of noise from mining plant. More mundane practices, such as enclosure of equipment and the fitting of silencers, helps to reduce the problem.

Subsidence

One impact exclusively associated with underground mining is the ground movement and surface deformation – collectively known as subsidence – caused by the collapse of abandoned workings. The consequences can be severe, although this depends upon the extent and use of affected land, and the form of movement which occurs, such as fissures or trough-shaped depressions. Typical effects include altered ground slopes and vertical land displacement, which change groundwater and drainage flows and can disrupt the gradients of roads, railways or pipelines, and create strain in buildings and other fixed structures, causing damage and eventual collapse. As these occur after mining has taken place, the prevention of subsidence was not a high priority with mine operators, and large areas of the older coalfields, particularly in Western Europe, Western USA and European Russia, have suffered severely. In the USA it is estimated that a quarter of its 8 million undermined acres are subject to subsidence, causing damage of $30 million per annum. Although the undermined acreage will rise to only 10.5 million by 2000, the area subject to subsidence is expected to double to 4 million acres.[14]

Extensive research has been undertaken into ameliorative measures, but the occurrence and severity of subsidence depends upon the interaction of many factors, including the mining method, form of the deposit, nature and thickness of the overlying strata, and soil composition. In practice, some degree of subsidence is accepted, but attempts are made to make it as even and predictable as possible. Pillars of unmined coal are left between worked-out panels, with both having finely calculated dimensions which – when geological conditions are taken into account – produce regular collapse and allow accurate forecasts of surface changes to be made. Where especially sensitive areas or buildings exist, larger pillars can be left beneath them, as will happen with Selby Abbey in the British National Coal Board (NCB) development in that area. Some prevention of subsidence may also be achieved by backfilling workings with mine waste, but this has fallen out of favour on health and production grounds. The partial nature of all these remedies is recognised in most coal mining countries by legislation requiring mine operators to compensate property owners for damage caused by subsidence.

New Mining Technologies

Present surface and underground mining technologies will remain predominant until the next century, although they will be used on a greater scale, with a correspondingly increased impact. New, or intensified, environmental problems may be caused by the introduction of hydraulic mining on a large scale, a trend towards large solid waste sites serving a number of mines (as will happen in some areas of the UK) and an increase in waste from a growing number of coal preparation plants, some using new methods.

The introduction of completely new underground mining technologies will bring both environmental benefits and costs, which are at
present difficult to quantify.[15] Those methods which convert coal *in situ*
into useful products will obviate the need to bring large quantities of
solid waste to the surface, and to treat and transport large amounts of
coal. However, large quantities of combustion gases, particulates, toxic
reagents and other pollutants may reach the surface, whilst unpredictable subsidence is likely to occur. Partial, or fully, automated
mining may avoid these problems, but at a likely cost of increased solid
waste disposal needs.

Conclusions

As one US study of coal mining has commented: 'the disturbance of the
land during mining and the generation of solid, liquid and airborne
pollutant or waste streams during preparation and transport are
inherent in the technologies'.[16] Their impact rises in proportion to the
scale of mining, and their effects are often cumulative. However, it is
clear that good mining practices can mitigate their severity, although
often with a considerable economic penalty and the sterilisation of
considerable quantities of coal, as when pillars are retained to reduce
subsidence or mining is discouraged in environmentally fragile zones.
Most countries now encourage such practices by legislative action, and
the severity of regulations can be expected to increase in the future.

Much of the opposition which faces new coal developments in the
Western world is caused by popular awareness of the damage caused by
coal mining in the past, even up to the early 1970s in some instances. The
misdeeds of the perpetrators have now returned to haunt the con-
temporary industry. In a similar way, the results of present industry
activities will be visible to all in several decades time, influencing the
judgement of public and political opinion on the new mine developments which are constantly required to maintain or increase coal
production. If the industry, in the West at least, has not lived up to its
much advertised promises of concern for the environment, its progress
in the next century may be placed in jeopardy.

COAL TRANSPORT

All modes of coal transport – rail, truck, barge, ship, conveyor belt and,
indirectly, electricity transmission lines – have environmental and

health impacts.[17] Most require considerable amounts of land, which may act as barriers to the movement of people or wildlife, upset ecological balances and affect water flows and quality. Slurry pipelines and canals also require large quantities of water as their transport medium, much of which is lost by 'exporting' or evaporation.[18] Most transport modes also create noise, have a significant visual impact and release coal dust into the atmosphere. Accidents and other hazards place transport employees and the public at risk: this may include adverse effects caused by microwave radiation from power lines, although definite proof is so far lacking.[19]

Many of these impacts from transport are concentrated in areas which derive little economic benefit from the production or use of the coal which is being moved, and which are especially concerned about environmental matters. Opposition to the operation of existing, and the construction of new, coal-transport facilities may be a significant constraint on increased coal utilisation in the future.

COAL USE

Coal Combustion

Combustion of coal produces a wide range of solid and gaseous substances, and may require large amounts of land and water, all of which may have deleterious effects on environment and health (see Table 3.3). Although new combustion technologies promise some improvement in environmental impact, existing technologies are likely to predominate until the end of the century.[20]

SULPHUR DIOXIDE

On combustion, the sulphur in coal is converted to sulphur dioxide, although a small proportion – usually less than 10 per cent – remains in the ash as sulphate compounds. Coal creates 40–70 per cent of man-made sulphur dioxide sources, which in total are considerably greater than those from natural sources. As high concentrations of the gas have adverse health and ecological effects, most countries either control the level of emissions, or require dispersal to lower ambient concentrations.[21]

Measures to achieve these objectives can be achieved before, during or after combustion. Before combustion, preference can be

(*continued on p. 82*)

TABLE 3.3 Some pollutants from coal combustion[a]

Substance[b]	Toxicity		Sources of pollution	US environmental standards[c]	Comments
	Acute	Chronic			
Gases Sulphur dioxide	Increased respiratory impairment – morbidity and mortality – in combination with particulates	Increased respiratory disease and decreased respiratory function with particulates	Sulphur contained in fossil fuels, smelters volcanoes	$365 \mu g/m^3$-24 hour max. $80 \mu g/m^3$- annual mean	Coal combustion presently represents between 60% and 70% of US SO_2 emissions
Nitrogen dioxide	Increase respiratory infections	Changes suspected in lung function; emphysema	Nitrogen fixation in high temperature combustion, and from nitrogen contained in fossil fuels: coal, oil, gasoline combustion	$100 \mu g/m^3$-annual mean	Organically bound fuel nitrogen is a more important component for coal NO emissions than for the other fossil fuels
Carbon monoxide	Behaviour changes, nausea, drowsiness, headaches, coma, death	Increase risk of coronary heart disease—arterial sclerosis suspected	Incomplete combustion of fossil fuels: coal, oil, gas, gasoline combustion	$40 mg/m^3$ 1 hr max. $10 mg/m^3$ 8 hr max.	

TABLE 3.3 Some pollutants from coal combustion^a (*Contd*)

Pollutant					
Ozone	Increased respiratory infection, eye irritation, headaches, chest pain, impaired pulmonary function	Unknown	Photochemical reactions involving hydrocarbons, nitrogen oxide and other compounds in lower atmosphere; reaction of atomic oxygen and oxygen in upper atmosphere	$160\,\mu g/m^3$ 1 hr max.	
Aromatic hydrocarbons	Fatigue, weakness, skin paresthesias (>100ppm)	Irritation, leukopenia and anaemia. Certain compounds are mutagens and carcinogens	A broad class of compounds naturally evolved from organic material, and from the evaporation and combustion of fossil fuels and other organic industrial chemicals	Non-methane HC $160\,\mu g/m^3$ 3 hour	Higher concentrations likely from less efficient and smaller boiler operation. Higher concentrations possible proximate to coal conversion facilities. Standard designed for photochemical oxidant control

TABLE 3.3 Some pollutants from coal combustion[a] (*Contd*)

| Substance[b] | Toxicity | | Sources of pollution | US environmental standards[c] | Comments |
	Acute	Chronic			
Particulates Total suspended particulates	With SO_2 in episode conditions contributes to mortality and morbidity	Pulmonary irritation, chronic obstructive and restrictive lung disease	Soil erosion, natural volcanoes and fires, industrial activity, fossil fuel combustion: coal and oil, secondary atmospheric conversion of gaseous compounds	$260 \mu/m^3$ 24 hour max. $75 \mu g/m^3$ annual ave.	A very broad class, undifferentiated by particle size or chemical composition
Sulphates	Increased respiratory disease; breathing difficulty in asthmatics	Respiratory disease and increased mortality suspected	Conversion of SO_2 to sulphates in the atmosphere, therefore primary sources are SO_2 emissions from coal and oil combustion. Smelters, kraft paper mills, sulphuric acid plants also produce sulphates. Na-		

TABLE 3.3 Some pollutants from coal combustion* (*Contd*)

			tural sources – H_2S emissions, volcanoes, sea salt	
Nitrates, nitrites	Increases infant susceptibility to lower respiratory infection due to conversion of nitrates to nitrites	May combine with amines to form carcinogenic nitrosamines, also mutagenic and teratogenic. Nitrites a direct animal carcinogen	Conversion of NO_2 to nitrates and nitrites in the atmosphere; therefore primary sources are NO emissions from fossil fuel combustion, fertiliser production, munition production, chemical plants, auto and industrial emissions	
Organic matter	Unknown for many compounds. Specific toxicity for others	Long-term is potentially carcinogenic and mutagenic	Fossil fuel direct and indirect use – combustion, refining, plastics, tars, coking, chemical production	Higher concentrations likely to be associated with non-direct combustion of fuels

TABLE 3.3 Some pollutants from coal combustion[a] (*Contd*)

Substance[b]	Toxicity		Sources of pollution	US environmental standards[c]	Comments
	Acute	*Chronic*			
Arsenic (oxide forms)	Effects large to small depending on form and route of exposure; rarely seen	Carcinogen, and teratogenic cumulative poison	Weathering; mining, and smelting; coal combustion; pesticides; detergents	0.05mg/l drinking water	
Beryllium	Short-term poison at high concentrations, especially toxic by inhalation	Long-term systemic poison at low concentrations; carcinogenic in experimental animals	Industrial; combustion of coal, rocket fuels	$0.01\,\mu g/m^3$ hazardous air pollutant	
Cadmium	Very toxic at high concentrations; to animals and aquatic life. Toxic by all routes of exposures	Possible carcinogen, cumulative poison; associated with hypertension, cardiovascular disease, kidney damage	Weathering; mining and smelting, especially of zinc; iron and steel industry; coal combustion; urban runoff; phosphate fertilisers	0.010mg/l drinking water $40\,\mu g/l/day$ proposed effluent standard (withdrawn)	Chronic cadmium poisoning resulting in illness and death has occurred in Japan, where cadmium mobilised by mining contaminated daily diet. Margin of safety – measured levels of cadmium

TABLE 3.3 Some pollutants from coal combustion[a] (*Contd*)

				in renal cortex compared to threshold for renal dysfunction – is low: 4 to 12.5
Chromium	Hexavalent form most harmful; skin and respiratory tract irritant	Carcinogenic; workers engaged in manufacture of chromium chemicals have incidence of lung cancer, no evidence of risk in non-occupational exposure	No chromium now mined in US. Emissions from industrial processes, including electroplating, tanning, dyes; coal combustion	0.05mg/l drinking water
Mercury	Methyl mercury and mercury fumes very toxic; other forms of variable toxicity	Methyl mercury very toxic, cumulative poison; affects central nervous system	Weathering; volcanoes; mining and smelting; industrial; pharmaceuticals; coal combustion	0.002mg/l drinking water; maximum of 2 300 grams mercury in emissions from	Environmental pollution leading to contamination of fish and shell fish caused illness and death in Japan;

TABLE 3.3   Some pollutants from coal combustion[a] (*Contd*)

Substance[b]	Toxicity		Sources of pollution	US environmental standards[c]	Comments
	Acute	*Chronic*			
			sewage sludge; urban runoff; fungicides	stationary sources; 20 µg/l/day proposed effluent standard (withdrawn)	contamination of fish in US has caused closure of waters to commercial fishing
Selenium	Soluble compounds are highly toxic	Probable carcinogen; also essential for life	Natural; mining and smelting; industrial process; coal combustion	0.01 mg/l drinking water	Interacts with other metals, increasing or decreasing toxicity

[a] This table is provided merely to indicate some of the substances which are potential environmental hazards along with some information on each regarding toxicity, sources, standards, etc. It is not to be interpreted as definitive.
[b] The substance listed is not necessarily the form in which it becomes a potential environmental threat. In some cases the oxide or some metabolite, rather than the substance itself, is the culprit.
[c] As of 1979.

SOURCE   Congress of the United States, Office of Technology Assessment, 'Direct Use of Coal' (1979).

TABLE 3.4 Sulphur reactions in the atmosphere

Reaction type	Reaction	Catalyst/special conditions	Comments
Indirect photo-oxidation	$SO_2 \xrightarrow[\text{radicals}]{\text{Free}} H_2SO_4$	Smog, sunlight, water vapour, HO, HO_2, CH_3O_2 radicals	Important reaction rates up to 5 per cent per hour giving half-life of SO_2 of $1/2 \rightarrow 2$ days. Rate depends on SO_2, hydrocarbons, NO_x levels, amount of sunlight
Heterogeneous catalytic oxidation	$SO_2 \xrightarrow{O_2} SO_4^=$	Liquid water; metal ions	Because of dependency upon catalyst concentration, probably important in plumes and polluted urban atmospheres; probably minor in rural atmospheres. Virtually zeroth order in SO_2.
Heterogeneous oxidation by strong oxidants	$SO_2 \xrightarrow[H_2O_2]{O_3} SO_4^=$	Water droplets	Importance in dispute, rate estimates vary by a factor of 100
Heterogeneous oxidation in the presence of ammonia	$SO_2 \xrightarrow{H_2O} H_2SO_4$ $NH_3 + H_2SO_4 \rightarrow NH_4^+ + SO_4^=$	Water droplets	Rates unknown, dependent upon availability of ammonia and pH of droplet
Surface catalysed reactions		Soot	Soot has been shown in the laboratory to catalyse oxidation of SO_2. Importance unknown

SOURCE Congress of the United States, Office of Technology Assessment, 'Direct use of coal' (1979).

given to coals of low sulphur content, either as the sole source of supply, or for blending with high-sulphur coals. Coal washing can also be used to remove up to 70 per cent of inorganic sulphur, although this requires substantial capital expenditure and energy use for subsequent drying. Disposal of the waste products creates further problems – as is also the case with the chemical or magnetic separation methods which may be introduced in the future. Coal can also be converted to a liquid or gaseous product, with sulphur and other substances largely removed, although this too is costly and has a great environmental impact.

Sulphur removal during coal combustion will be achieved by the introduction of fluidised-bed boilers, in which limestone or dolomite reacts with the sulphur dioxide to form calcium sulphate and sulphide. However, these will require large quantities of raw materials – the extraction of which itself causes environmental problems – and produce a contaminated sorbet with high disposal costs. At a more general level, emissions from coal combustion can also be reduced by giving priority in an electricity system merit order to low sulphur dioxide-creating boilers, where this option is possible.

Until recently, the most prevalent method of sulphur dioxide control has been the tall chimney stack, which vents emission gases above ground level. These have been successful in preventing high local concentrations, although this may be partially at the expense of more distant regions. As no other control technologies are 100 per cent effective, stack dispersal will always be an important way of mitigating the effect of sulphur dioxide and other emissions.

In some countries, notably the USA, the installation of flue-gas desulphurisation (FGD) equipment is mandatory for new, large coal-fired boilers. These may be either 'throwaway', that is, producing an essentially valueless product which must be disposed of as waste, or 'regenerable' technologies, with the latter regenerating the agent used to capture sulphur dioxide, and providing other valuable products to defray process costs.[22] A further distinction is made between 'dry' and 'wet' technologies, with the latter bringing the gases into contact with a slurry or solution. Wet processes typically remove up to 90 per cent of sulphur dioxide, and often considerable quantities of particulate matter. Their operation is impeded by scaling and corrosion of equipment, while the cooling of flue gases leads to condensation in the stack or atmosphere, and a reduction in plume height and transparency. Dry processes are less developed, but are likely to have equivalent disadvantages, although these will be offset by the fact that they do not produce large quantities of sludge for disposal.

Most existing desulphurisation technologies are based on lime/ limestone slurries, although the dual-alkali process uses a solution of sodium and calcium alkalis. All produce a sludge of up to 40 per cent solid matter, mainly in the form of calcium sulphite and sulphate, which is either stored in lagoons or used as landfill. Its stability can be increased by blending with fly ash, or the addition of hardeners and other chemicals, and calcium sulphite can be oxidised to sulphate, which is easier to dewater and handle, and can in some areas be a saleable product. Despite this treatment, problems can be caused by seepage or leaching of toxic components which are costly to prevent, especially when the sludge is classified as a hazardous waste.

The scale of these problems will be reduced by the development of regenerable wet processes, of which those based on sodium sulphite (Wellman-Lord) and magnesium oxide are the most advanced. All these convert sulphur dioxide into saleable sulphur or sulphuric acid, although there is the possibility that large-scale use would glut the market and lower prices, adversely affecting their economics. Storage of such products if they were unable to be sold would also create new hazards. Other problems of these technologies include high-energy usage and operating complexity.

The development of new desulphurisation technologies will give greater flexibility to utilities and other coal users. Many will operate two processes in series, with a cheap, low-efficiency unit removing the bulk of sulphur dioxide and a more expensive, higher efficiency unit the remainder. Others will integrate them with coal preparation and combustion technologies into a sulphur control system. Nevertheless, the thermal efficiency of coal use will be reduced and substantial costs will be incurred. One US study has estimated desulphurisation adds 20 per cent to the capital cost and 100 per cent to the non-fuel operating costs of a standard coal-fired power plant, with a 5 per cent reduction in its generating capacity.[23] Other studies have suggested that to reduce sulphur-dioxide emissions by half would cost $5 billion per annum in the USA, and $10 billion per annum in Western Europe, although these might be balanced by a similar level of benefits.[24] However, considerable uncertainty surrounds all such estimates.

NITROUS OXIDES

Coal combustion creates nitric oxide (NO) and nitrogen dioxide (NO_2), with the former being oxidised to the latter when sufficient oxygen is present. They are usually measured together as nitrous oxides and typically amount to 25–40 per cent of sulphur-dioxide emission levels. Their main source is the oxidisation of organic nitrogen compounds within the coal – which typically contains 1–1.4 per cent nitrogen – although quantities are created by reaction of nitrogen and oxygen in combustion air at sufficiently high temperatures.

Coal-derived nitrous oxides account for 20–30 per cent of those from man-made sources, which amount to only 10 per cent of the total from natural sources. However, their concentration – particularly in urban areas with special atmospheric conditions – has led to emission controls in some countries, a trend which is likely to spread in the future. Present control methods are designed to reduce nitrous-oxide formation, by lowering flame temperatures and/or the availability of oxygen.[25] The first objective is usually achieved by burner redesign or recirculation of flue gases into the combustion zone, and the second by two-stage combustion, in which initial coal combustion occurs in oxygen-deficient conditions. These techniques can reduce nitrous-oxide emissions by up to 70 per cent, although often at the expense of thermal efficiency and an increase in particulate levels as a result of incomplete combustion.

In some cases, removal of nitrous oxides from flue gases will also be necessary. One process uses a gaseous reagent, usually ammonia, to reduce them to elemental nitrogen, sometimes by injection of the reagent into the boiler itself. Another process uses a liquid reagent, first to absorb and then to reduce the nitrogen. Both are likely to be costly, create operational difficulties and require disposal of waste products.

OTHER GASEOUS EMISSIONS

Coal contains chlorine, which is converted to chloride compounds and hydrochloric acid gas (HCl). Most chlorides are found in the ash, but some vaporise to cause severe corrosion problems in combustion equipment. This is also the case with hydrochloric acid gas, which contributes to flue-gas acidity. In some countries, coal is blended to reduce its chlorine content, and in the UK some coal reserves are not exploited because of their high chlorine level. Carbon monoxide may also be produced by incomplete combustion of coal. Many organic compounds, trace elements and radionucleides are also vapourised in coal combustion, although most condense as temperatures fall. They are often disproportionately concentrated on the fine particles which are most adverse to human health. Particular problems occur with arsenic, selenium, heavy metals and polycyclic organic matter (POM) such as benzo(a)pyrene and dibenzo(a,h)anthracene.

ASH AND PARTICULATES

Coal combustion creates large quantities of ash, of which up to 40 per cent remains within the combustion chamber, usually as a semi-fused slag (bottom ash), with the remaining, lighter material (fly ash) entrained in the combustion gases. Small quantities of additional particulate matter are also created by reactions between components of these gases.

The visual intrusion of particulate emissions, and their effects on human health, have produced long-standing regulations in most coal-using countries and control technologies are therefore well developed.[26] In the USA in 1978, 92 per cent of the fly ash generated in large coal-fired boilers was captured, and present regulations there and elsewhere require 99 per cent plus capture for new coal-fired boilers.

Four particulate-control technologies are presently available. In electrostatic precipitation, particulates within flue gases are electrically charged and attracted to a collection plate of opposite charge. This is highly effective, but has high capital and operating costs, especially for low-sulphur coals. Mechanical collectors use gravity, inertia or centrifugal forces to capture particulates. They are less effective than other methods, but cheap to operate, and are often installed as the first unit in dual-stage systems. Wet scrubbers wash particulates from flue gases by water sprays, and are generally less effective and more expensive than other methods, although they may be installed as an element in sulphur dioxide-control equipment. All the previous systems share the disadvantage of achieving better results with larger particulates, which are less damaging to human health than smaller ones. The most suitable technology for capturing these fine particulates is the fabric filter baghouse, although its performance is steadily reduced by components of the flue gases and it is expensive to operate. Despite this, their use is likely to greatly increase in the future.

Most coal ash is in the form of chemically-inert oxides, particularly of silicon, aluminium and iron, but sulphates, trace elements and radionucleides are also present[27] (see Tables 1.3 and 1.4). Most of these – including chromium, cobalt, manganese, strontium and thorium – are unvolatised and distributed between bottom and fly ash in relatively low concentrations. Others – including arsenic, beryllium, cadmium, copper, lead, selenium, uranium and zinc – are volatised, and subsequently condense on to fly ash particles, allowing high concentrations to occur in some instances. The greater surface area-to-volume ratio of fine particulates mean that these absorb a disproportionate quantity of both these substances and others which are volatised, particularly organic compounds.

The particulates from coal combustion which eventually reach the atmosphere form approximately a third of those from man-made sources, which in total are less than those from natural sources such as volcanoes. As a result of the extensive use of pulverised-fuel boilers, particulates from coal combustion are often smaller than those from other sources. Little account has been taken of this fact when emission regulations have been set, and in some countries a reduction in overall coal particulate emissions has been accompanied by an increase in those of fine particulates. Measures aimed at the latter are likely to be introduced in many countries during the 1980s, although the problem is likely to be a source of concern for some years.

LAND AND WATER IMPACTS

Coal use requires considerable amounts of land, not only for the siting of combustion, turbine and other equipment, but also for storage of incoming coal and disposal of waste products. The total needs of a 1 000MWe power plant over its operating lifetime may be up to 1 000acres.[28] In many countries the large blocks of land required are scarce and highly valued for alternative purposes, and public opposition to new developments may be considerable. One consequence of this may be more intensive use of existing sites by adding new coal-using facilities to them.

The land required for disposal of ash and, increasingly in the USA, scrubber sludge – which takes up to five times as much space as an equivalent weight of ash – is a particularly controversial aspect of coal use. Although considerable quantities of ash can be sold as construction material, or used in some industrial processes such as portland cement manufacture, most has to be permanently disposed of, usually by landfill. In many cases, the ash is slurried to improve handling, requiring the use of settlement lagoons, which are also necessary for sludge disposal. During these processes, large quantities of dust may be created, and seepage or leaching of toxic substances into water supplies may occur. In the USA it has been estimated that over 100 000 acres of land will be required for ash and sludge disposal by 2000.[29]

Coal use for electricity generation also has considerable effects on water quality and availability, stemming from the use of cold water to cool and condense steam at the end of the turbine cycle. In open-cycle cooling systems, water is taken from a large source such as rivers, lakes or the sea, and subsequently returned to it at a much higher temperature.[30] Although this minimises water losses, changes in salinity and dissolved oxygen and solid content occur, with deleterious effects on aquatic ecology. Organisms may also be directly affected by the increased temperatures, or by being sucked with the water into the cooling system itself.

In closed-cycle cooling systems, water is taken from similar sources as open-cycle, but allowed to evaporate and escape as steam to the atmosphere, usually through large, visually-intrusive cooling towers. These reduce – but do not entirely remove – the problem of thermal pollution in surface water, but at the cost of much greater water losses. The unevaporated water which is returned to its source may also contain large concentrations of mineral salts and the biocides used to prevent fouling of tower interiors. The release of large quantities of steam may also have microclimatic effects in the area around the plant.

In the future, air-cooled 'dry' systems may be introduced, either alone or with supplementation by evaporative systems in summer, when the cooling capacity of ambient air is reduced. These will mitigate the water losses and thermal pollution of present systems, but will have a greater capital cost and reduce thermal efficiency.

COAL POLLUTANTS IN THE ATMOSPHERE

Those coal pollutants which reach the atmosphere are subject to complex, and little understood, chemical reactions and the effects of weather, resulting in their return to earth as rain or snow[31] (see Table 3.4 on p. 81). Although these changes occur quickly – sulphur dioxide has an average residence time in the atmosphere of one day – local concentrations can be high, causing health, ecological and property damage. Research has also demonstrated that, under favourable conditions, coal-derived pollutants can be transported long distances. In Europe, 20–60 per cent of sulphur-dioxide emissions from within the European Economic Community (EEC) are exported to other countries.[32] An emission plume from a St Louis, Missouri, power station was found to be relatively undispersed at a distance of 180 miles, and satellites have tracked 'hazy blobs' of polluted air across the USA, showing their tendency to be transported by prevailing winds to the east coast.[33] Other

factors which determine the extent of atmospheric transport include
terrain; vertical temperature variations; cloud conditions and the
movement of high- and low-pressure systems.

High concentrations of coal pollutants can create considerable
ecological problems, although these will vary according to atmospheric,
soil moisture and other conditions.[34] Research has shown that many
species are sensitive to sulphur dioxide, even at low concentrations.[35]
Exposure as small as 0.02–0.05ppm can affect the growth and commercial yield of crops such as alfalfa and ryegrass, although the extent of the
damage depends upon wind speed – a crucial factor in determining
sulphur dioxide uptake – and the presence of other environmental
factors which affect growth. Lower plant species, such as ferns and
mosses, are even more sensitive. High concentrations of sulphur
dioxide – which often occur when power-station plumes fall to the
ground – produce a characteristic burning of foliage, with marked
effects on plant vitality. As with humans, damage is increased when high
concentrations of other pollutants, especially ozone, nitrous oxides and
oxidants, and hydrofluoric acid, are also present. Similar effects have
been found for sulphates, although neither these nor sulphur dioxide
appear to have major effects on animals. Experiments have also shown
that these substances can encourage growth in areas of sulphur-deficient
soils and that some grass species have developed a resistance to them.

Nitrous oxides appear to be less harmful than sulphur dioxide, and
levels from coal combustion are only a fraction of those from natural or
other sources. However, photochemical oxidants, particularly ozone,
cause considerable damage, which has been charted in California and
the Western USA (a particularly vulnerable area).[36] Crops and native
species have suffered foliage burning, reduced photosynthesis and
diminished growth. Although animals are not directly affected, they can
be hit by loss of cover and food sources.

The worst effects of sulphur dioxide and nitrous oxides relate to the
formation of 'acid rain', one of the most severe environmental problems
in the northern hemisphere.[37] This occurs when these emitted gases
react in the atmosphere to create sulphuric and nitric acids, which are
then dissolved to reach the earth in rainwater or snow. Although the
effects of this are partially buffered by alkaloid compounds in the
atmosphere, in North America and Europe the pH (acidity) value of
rainfall has been steadily falling. One Scottish storm of 1974 produced
rain with a pH of 2.4 – equivalent to strong vinegar.

Acid rain corrodes buildings and other man-made objects – it is
thought to be a prime cause in the deterioration of the Colosseum in

Rome and the Parthenon in Athens – and produces damage estimated at billions of dollars. By lowering the pH of lakes and watercourses, it kills many species of fish – trout and salmon die when the pH falls below 5 – and then other aquatic organisms, such as algae and plankton. Spring is often the most destructive time, for melting snow can produce a sudden surge of highly acidic water which has been stored through the winter.

In the Adirondack mountains of New York, USA, pH levels in upland lakes fell from 6.5 in the 1930s to 4.8 in 1975 and over 90 per cent of these waterbodies no longer have fish populations. In Sweden, the Ministry of Agriculture has estimated that 20 000 of the country's 100 000 lakes have been seriously affected, and in the worst cases has organised the dumping of lime into them to neutralise the acidity.

Acid rain also affects land organisms, especially where the soil is already acidic, as in the pine forests of Northern latitudes. It kills the bacteria which recycle dead matter, increases the leaching of nutrients from the soil, and renders toxic metals, such as lead and mercury, more easily absorbed by living creatures. Partially as a result, shrubs and other vegetation die, encouraging soil erosion and reducing the yield of forest lands. Two areas which have been especially affected are Eastern USA and Canada, and Scandinavia, which have been subject to both their own emissions and those from other regions or countries. Political or diplomatic pressure may result in the introduction of stringent controls in source areas, affecting the economics of coal use.

CARBON DIOXIDE

One of the main products of coal combustion is the harmless gas, carbon dioxide.[38] Continuing emissions from coal and other man-made and natural sources has produced an increase in atmospheric carbon dioxide levels from an estimated 290ppm in 1900 to about 330ppm in 1979. Most observers agree that this increase will continue, with upper forecasts suggesting that the level will be 360–400ppm by 2000, and 800ppm by 2040.

This trend caused concern during the 1970s because of the 'greenhouse effect' it may create. Carbon dioxide transmits visible light but absorbs certain infra-red wavelengths, hence trapping some of the heat which would otherwise have radiated from the earth. By some estimates, a doubling of atmospheric carbon dioxide levels would produce a 2–3°C rise in global mean surface temperature. This would be disproportionately concentrated at the polar regions, producing melting of icecaps, increases in sea levels and climatic changes. Scientists have modelled the climatic variations which might occur, based on past changes in rainfall and temperature when the earth underwent warmer periods in the past.[39] This suggests that most regions would experience temperature increases, that precipitation would increase markedly in Canada and India, and decline in much of the USA, Europe and USSR.

However, some scientists are doubtful that such increases or effects will occur. Human activities in the last century have released more carbon dioxide that has been retained in the atmosphere, with the excess stored in carbon 'sinks' such as biomass and ocean waters. Although there is no certainty that this will occur in the future, optimists argue that atmospheric feedback mechanisms exist, which automatically dampen the rise in carbon dioxide levels. In particular, they believe that increases in atmospheric carbon stimulates photosynthesis in plants, and that rising temperatures increase evaporation and cloud cover, reducing the amount of solar radiation which reaches the earth's surface. It is also true that the relationship between increases in carbon dioxide levels and temperature rises on the earth is still unknown, so that large increases in the former could have only limited effects on the latter.

Pessimists, however, argue that positive feedback might occur, with increased cloud cover enchancing the 'greenhouse effect'. Uncertainty about future carbon dioxide levels is matched by that about its present and past sources, and its interaction with other climatic factors. It is known that burning oil produces only 80 per cent, and gas 57 per cent, as much carbon dioxide as coal – which typically create 200 pounds of carbon dioxide per million Btu – and that coal-derived synfuels produce up to 40 per cent more.

However, the relative proportions coming from fossil fuels or other sources, notably the destruction of tropical rainforest, is unknown. Many scientists believe that the latter is a more pressing problem than any other, but coal combustion may be the easiest source of carbon dioxide to be controlled, if this was necessary. Finally, there are other man-made factors which may also cause climatic change, several of which are associated with coal combustion. They include changes in the earth's albedo and cloud cover by atmospheric dust; increased cloud cover from greater evaporation on the earth, for example from irrigation schemes; and thermal pollution of the atmosphere. Some of these, together with natural factors – such as atmospheric dust from volcanic eruptions – may act to lower the earth's temperature by screening out sunlight, therefore counter-balancing the greenhouse effect.

There is no doubt that any increase in temperature caused by rising carbon dioxide levels could have major effects on global climate and economy. However, the consequences would not become evident for many years, and present research will not produce definite conclusions for some time. Even then, it may not immediately affect coal production. As the example of acid rain shows, knowledge is a necessary, but not sufficient, condition for a problem to be solved. Nevertheless, it may well be that a relationship between coal combustion and climatic change will dictate a policy of prudence in many countries, particuarly towards the end of the century, and that more optimistic forecasts of future production may not be met. This is particularly likely if the unusual weather patterns of the 1970s were to be repeated in the 1980s.

Finally, on a microclimatic level, the possible effects of flue-gas, ionised particle and thermal discharges on precipitation and sunlight may also create some local opposition to coal use.

CONCLUSIONS

Coal combustion creates many enviromental and health impacts, which can be reduced but never eradicated. Control measures have substantial economic and energy efficiency costs – although these may be offset by benefits in the rest of the economy – and it is clear that they can act as a disincentive to coal development, as partially happened in the USA during the 1970s. New combustion technologies, such as fluidised beds and MHD power units, or better coal-preparation methods, may give some improvement, although neither will be widespread until the end of the century.

These facts, and the presently inadequate knowledge of the effects of many coal-combustion products, are likely to make coal use a more, rather than less, controversial subject as its scale increases in coming decades. To many – especially those who do not appear to gain any direct benefit from coal expansion – the adverse impacts of coal will seem unacceptable, and either more stringent environmental controls or the use of other energy sources will be canvassed. But such beliefs will face powerful opposition from energy policy-makers, who feel that such controls will undermine the competitiveness of coal, especially for electricity generation, place a heavy burden on national economies and perpetuate Western dependence on Middle Eastern oil.

COAL CONVERSION

Severe environmental and health problems have been associated with coke- and gas-making plants in the past. Although great improvements have been made in recent decades, it is likely that the size and similar nature of chemical processes associated with the new generation of coal conversion plants will also create hazards in the working and external environment.[40]

Much of the impact of conversion plants will be similar to that of electricity power plants, especially in terms of land and water requirements, and noise or visual intrusion. In some cases, coal conversion plants may be more benign than power plants due to their high energy efficiency – 60–70 per cent compared to the 35 per cent of the best coal-fired power plants – so that less pollutants are released per unit of energy gained. Process characteristics also aid the capture of some harmful substances, such as sulphur and nitrous oxides, which are presently associated with coal combustion.

However, these advantages may be obviated by the wide range of organic and inorganic compounds which are produced during coal processing, many of which are known to be carcinogenic, mutagenic or toxic. Particular problems may be caused by the polycyclic aromatic hydrocarbons (PAH), such as benzo (a) pyrene, dibenzo (a, h) anthracene, and 3-methylcholanthrene, which were the main cause of the high cancer rates of gas and coke workers in the past. Liquid streams exiting from conversion plants are also likely to contain high concentrations of phenols, cyanides and other hazardous organic compounds although biological and mechanical processes have been developed which may be suitable for their treatment.[41] Waste gas streams from conversion plants may contain hazardous concentrations of organic material, hydrogen sulphide, ammonia and carbon monoxide. Pollution controls may be less effective during certain periods, as in start-up or shut-down, or the even of explosion or fire within the plant. Accidental releases may also occur, as when 2–3000 gallons of waste water contaminated soil and groundwater at an SRC-2 liquefaction plant at Fort Lewis, Washington. Such problems would be especially likely in the case of explosion, which has occasionally happened in conversion plants. Too little is known about conversion processes to form definite conclusions but, as a US study has commented, they 'may pose serious problems for occupational health and safety and accidental public exposure' and 'have significant enviromental constraints that will require extensive research and development before they can be widely used or before they can be deployed in some enviromentally sensitive areas'.[42] Such doubts mean that pollution-control costs will be a major part of the expenditure of conversion plants. Intense public suspicion can also be expected – prefigured by local opposition to the siting of the first synfuel demonstration plants in the USA during 1980 – and any evidence of deleterious health and environmental consequences will threaten the future of the industry.

COAL, HEALTH AND SAFETY

As with every human activity, the production and use of coal has a cost in terms of the deaths, injuries or illness which it causes.[43] Although many preventative measures have been introduced during the present century, with considerable success, these costs remain substantial and are an impediment to coal expansion. They are, however, difficult to quantify and are unevenly distributed within society, although a basic distinction can be made between occupational and public hazards.

Occupational Hazards

The main hazards of underground mining are respiratory diseases caused by high concentrations of coal and rock dust, especially chronic bronchitis and pneumoconiosis, and accidental injuries or deaths.[44] The risks of contracting some types of cancer may also be increased, although for other types – including lung cancer – it may be reduced. In general, mechanisation has reduced the levels of deaths and injuries – which are now caused mainly by underground transport equipment – but has increased dust formation, although the potentially adverse health consequences of this have been counteracted by improved protective measures. The level of risk varies according to mine characteristics, with one US study finding that small mines have twice the accidental death rates of large, and that if the standards observed in all mines were brought up to the safest accident rates could be cut by up to 75 per cent.[45] The underground mining of coal is also more hazardous than that of other minerals – with, in the USA, up to 50 per cent more injuries of greater average severity. Comparisons with manufacturing industry are even more marked with, in the USA, 0.32 fatal accidents for coal mining against 0.03 in manufacturing, per 1 000 man-years. Equivalent figures for the UK are 0.27 in coal and 0.03 for manufacturing, and for Czechoslovakia 0.27 and 0.07.[46] These figures underestimate the actual hazardousness of underground coal mining by including figures for surface mining, which imposes similar degrees of accidental risk on its employees as other industrial occupations, and causes fewer deaths from respiratory disease.

A large number of accidents and injuries occur in connection with the surface transport of coal and coal products, although these have been reduced by new operating methods, particularly the introduction of unit trains. Similar improvements have reduced the hazards associated with coal processing, particularly those caused by mutagenic and carcinogenic substances. These were first observed by Percival Pott in 1775, who noted a high incidence of cancer of the scrotum amongst chimney sweeps, and were confirmed in the twentieth century by several epidemiological surveys in gas and coke plants.[47] They may pose a severe problem in the new generation of coal liquefaction and gasification facilities. Substantial hazards may also be created by the storage, handling and distribution of coal-derived products, especially coal liquids, which generally contain higher concentrations of carcinogens and toxins than their petroleum equivalents.

When the occupational risks of coal-based energy systems are calculated, respiratory diseases and accidents are by far the greatest hazards. The greater efficiency of coal conversion or direct combustion routes means that these are less hazardous than coal – electricity routes.

Public Hazards

Coal mining has a significant, but relatively minor, impact on public health and safety. Particular hazards are caused by the collapse of waste tips or tailing dams; releases of dust into the atmosphere and the leaching of toxic materials from mine wastes into water supplies. Some accidental deaths and injuries are caused by coal transport, which also causes some problems by the release of coal dust.

Concentrations of coal pollutants, particularly in combination, cause considerable damage to human health, although the extent of this is uncertain.[48]

Until recently, most attention has been focused on the hazards of particulate emission from coal use, both because of their effects on health – particularly respiratory diseases – and their visibility. Although problems have been mitigated by the great reduction in particulate emissions which has occurred in most coal-using countries during the last thirty years, concern is now centred on fine particulates (those less than 2–3 μm in diameter), emissions of which have increased as a result of pulverised-fuel combustion methods and the ineffectiveness of existing control technologies. These can directly cause pulmonary irritation and lung diseases, and indirect effects as a result of the toxic, radioactive or carcinogenic substances they contain being leached into the bloodstream, or being absorbed in food.[49] Although the few studies which have been carried out on the latter hazard suggest that the risk is not unacceptably high – one found that 2–5 cases of cancer and genetic damage is caused by every thousand hours of power-plant operation – more stringent regulations may be introduced in the future, especially as some studies suggest that, in normal operation, radiation hazards from coal-fired power stations may be of a similar order to those from nuclear power plants.[50]

Typical concentrations of sulphur dioxide from coal combustion do not appear to be hazardous on their own, but can increase respiratory impairments and disease when particulate matter is present. Sulphates and sulphuric acid aerosols derived from sulphur dioxide are also thought to aggravate or cause respiratory diseases and to irritate the eyes and other organs. One US study, using sulphur-dioxide emissions as an index of the impact of all these pollutants, found that each 1 000 tons emitted caused 0–0.68 deaths and 0–980 cases of asthma aggravation amongst the public.[51] However, the use of zero as a lower value reflects the uncertainty of such assessments, and much of the earlier research in this field has been severely criticised.

Nitrous oxides are more reactive than sulphur dioxide, so that their effects are more localised. Research findings are ambiguous as to their direct effects on health, although some suggest that they impair respiratory functions and metabolic rates. More damage may be caused by the products of photochemical reactions in the atmosphere, particularly nitrates, nitrites and ozone.[52]

An overall assessment of the health impact of air pollution from coal combustion by a US study group concluded that the epidemiological evidence is inconclusive and that 'the present regulatory strategy must be considered as conservative in the light of existing evidence, but probably justified in view of the prospective rapid growth in the use of coal'.[53] Another US survey of all coal pollutants has stated that 'there are grounds for speculation . . . and some positive but not conclusive statistical evidence that current and expected future concentrations of coal-related pollutants may be dangerous to human health'.[54]

THE RISKS OF COAL

Assessing the risks of deaths, injuries and other adverse consequences from coal or other energy systems is difficult, and often impossible. The necessary data may not be available, some risks may have more certainty attached to them than others and in many cases subjective and statistical assessments may be at variance. As one study has commented, 'it is unlikely that the appraisal of risk will ever be able to avoid relative value judgements between different kinds of risks, as well as between risks and economic or other benefits of energy technologies'.[55] Nevertheless, such appraisals will continue, both because of political and social pressures for non-economic comparisons between energy systems, and as an element in finding the optimal balance between spending on preventative measures and the reduction in risk which this achieves.

At present, and with the caveat that much relevant information is not available or is of questionable validity, it appears that coal-energy systems have, in routine operation, a greater risk of causing health and environmental damage than other fossil fuel or nuclear routes.[56] This is particularly true of coal–electricity routes, although the nature of the risk depends upon the size of plant. Large plants create a greater environmental impact than a number of small plants, but have a better health and safety record.[57] However, there is probably more chance of substantially reducing the risks of coal – by greater mechanisation, more surface mining and safer operating procedures in coal production, new methods of coal transport and the introduction of new coal-using technologies – than is the case with other energy systems. Coal also suffers because its long history of use has allowed its risks to be assessed with a greater degree of certainty.

By comparison, one of the central issues determining the risks of nuclear power – the biological effects of low-level radiation – remains controversial and will only be fully resolved when long-term epidemiological studies have been completed.[58] Equally, the certain and persistent nature of routine coal risks must be balanced against the fact that the worst conceivable accident or incident associated with coal is much less catastrophic than a loss-of-coolant or other accident in a nuclear reactor. Although these are statistically unlikely, if they did occur many thousands of people would die and large areas of land would be sterilised for centuries, with the additional possibility of long-term genetic damage.[59] Finally, many of the risks associated with coal production and use are reversible, unlike those relating to radioactive material, which may persist for many centuries.

At present, a combination of inadequate information, opposition to other energy sources, the persistence of traditionally favourable attitudes to coal production and use in many regions and, in some countries, the lack of opportunities to dissent, means that the risks of coal are generally judged to be 'acceptable', compared to the benefits obtained. However, the dissemination of information, and the likelihood of a rapid build-up of coal production and use in many countries may change this situation. As one US study has commented: 'We must be prepared for the possibility that adverse health effects, global CO_2 increase and associated climatic change, freshwater supply problems, and ecological considerations will eventually severely restrict continuing expansion of coal use'.[60] For the USA, this study put the upper limit of acceptable risk at three times present coal output, a figure which may in time be revised downwards, both there and in other parts of the world.

CONCLUSIONS

The question of energy, environment and health became a crucial issue in the Western world during the 1970s. A number of factors – increased energy use; replacement of cheap Middle Eastern oil by other energy sources which will usually, have a greater environmental impact; a greater sophistication in monitoring adverse consequences; increased concentration of energy-producing and -consuming units and the existence of well-organised environmental movements in many countries – suggest that this will continue to be the case in the future.

In routine operation, present coal-energy systems are probably more hazardous than others but, in many cases, more economic or politically acceptable. The risks attached to coal production and use could be reduced if present best operating practices became the norm or if more stringent regulations were introduced, although with an economic and energy efficiency penalty. They are also more predictable than those of other energy sources, although there is the possibility that beyond a certain level of coal utilisation, qualitative changes in impact will occur.

The political support for increased coal utilisation which developed in the late 1970s and early 1980s, and the intense opposition to use of nuclear power from large minorities in many countries, means that environmental factors are not likely to prevent major coal expansion during the 1980s, although they may reduce its scale. Beyond 1990, however, the picture is more clouded. By then the level of the impacts from coal expansion will be clearer and, if the industry and governments have not lived up to their promises, public opinion may change rapidly. By this time, improved versions of present, or new, energy technologies may appear to be more environmentally benign than coal and be favoured at the latter's expense.

Increased concern at the impact of coal may be reflected in more stringent regulations. It may also result in inter-regional or -national conflicts between those areas or countries which appear to be net beneficiaries from increased coal utilisation and those which do not. The complexities of coal-energy systems – requiring co-ordination of mining, transport and user facilities, most of which have long lead times – mean that delays in one link, caused by such sectional opposition, can have damaging effects on others. Such delays are commonplace in the many Western countries with decentralised systems of government, especially those with Federal and state governments – as in three of the world's leading coal exporters, the USA, Australia and Canada. The USA, in particular, is noted for its openness to environmental and other interest groups, resulting in delays of several years to some energy projects. Problems of this type which are already apparent in the USA, and may intensify both there and in other countries, include: a reluctance of sparsely-populated and/or agricultural states to allow large-scale surface mining of coal, or the export of water in slurry pipelines; opposition of some communities or areas to increased rail traffic and the construction of new rail or port facilities; local opposition to the building of coal-using facilities or the dumping of solid or other wastes in areas remote from the markets they serve, balanced by opposition to the construction of coal-using facilities in urban areas which are already heavily polluted.

The transition of coal-energy systems from a national to a global basis, in terms of both economic organisation and environmental importance, will also increase the international importance of coal-environmental issues. In particular, the question of acid rain and, if the pessimistic forecasts are accurate, carbon dioxide build-up can be expected to become contentious diplomatic issues in the future.

4 The USA: Saudi Arabia of Coal?

In the 1970s the US coal industry began to recover from the depression which, though not as traumatic as in other countries, affected it from the 1950s on.[1] Indeed, after 1973 it was given great encouragement by successive administrations. After initially disappointing results production reached 824mt in 1980 but fell back to 815mt in 1981, to 820mst (744mt) in 1982.[2] Forecasts of future output are highly optimistic, based on a greater use of coal in the domestic energy system and an increase in exports. With recoverable reserves estimated at 191btce, and additional resources of 2 519btce – which, some observers believe, could make the USA the 'Saudi Arabia' of coal – there is no physical reason why production should not increase by large increments.

GEOGRAPHY

According to the US Geological Survey: 'Coal-bearing rocks underlie about 13 per cent of the land area of the 50 United States and about 14 per cent of the land area of the 48 coterminous states'.[3] In considering these resources, a broad distinction is usually made between deposits located to the East and West of the Mississippi River. Eastern coals are usually high-ranking, of high sulphur content and deeply buried. Western coals were generally formed about 150 million years later and are therefore lower in rank and nearer to the surface, whilst their sulphur content is generally low. Over 80 per cent of US resources are thought to be located in the West, but recoverable reserves are more evenly distributed. However, several studies have demonstrated the limitations of existing resource and reserve information, and there are probably large volumes of coal in low 'resource' categories which can be mined at lower cost than some deposits in the US Geological Survey's 'demonstrated reserve base'.[4] The main coalfields (see Fig. 4.1) are:

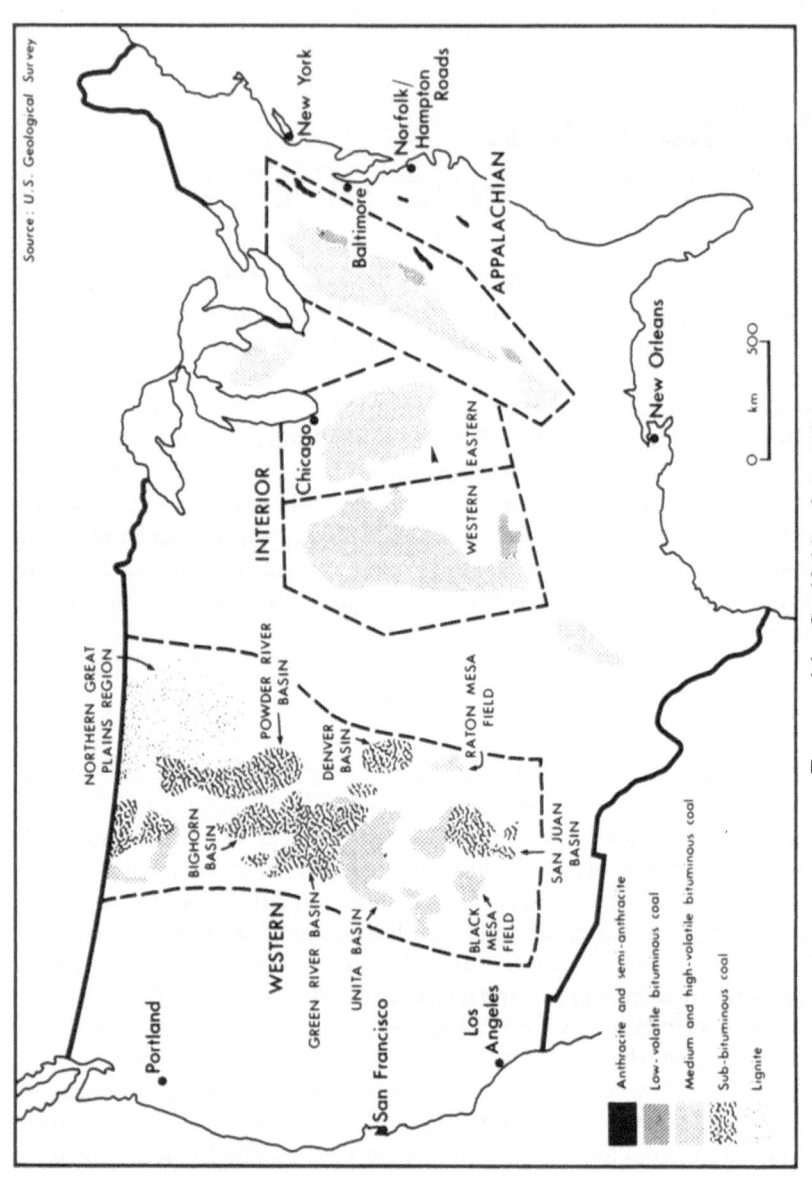

FIGURE 4.1 Coalfields of the USA

Appalachian Basin   This was the first field to be mined in the USA and covers an area of 180 000 square kilometres, from Pennsylvania to Alabama, and largely coterminous with the Appalachian Mountains.[5] Deposits are mainly high-grade bituminous, with medium ash and high sulphur content, although large anthracite deposits are found in Pennsylvania. Particular seams often continue over large distances, as with the famous Pittsburgh bed which covers an area of several thousand square kilometres and produced 59mt of coal in 1977. However, seam thicknesses rarely exceed two metres, and this degree of thinness, together with the steep terrain of the region, makes mining difficult. Most production is from underground facilities (which are relatively shallow by European standards), although an increasing percentage has been won by surface mining. In both types of production, a large number of mines of widely varying size exist. Surface mineable reserves are probably insufficient to support even present production rates for more than a decade, and the proportion of output from underground mines will increase rapidly in the 1990s.

Although the Appalachian Basin only contains a quarter of US recoverable reserves, it accounts for over half of national output, mainly from West Virginia, East Kentucky, Pennsylvania, Ohio and Virginia. The largest market for its coal is for metallurgical uses, with power generation a close second. The Basin is also the source of almost all US coal exports. However, production from most mines in the 1970s has been below capacity, causing financial problems and unemployment amongst miners. The reasons for this have included a failure of coal consumption to rise as rapidly as predicted, depression in the steel industry and the preference of utilities, particularly in Ohio, for low-sulphur Western coal. Although over-capacity is expected to persist for most of the 1980s, new environmental measures to inhibit utility preference for low-sulphur coal, and the increase in coal exports, should encourage moderate increases in production. This may be accompanied by changes within the Basin, such as a movement to the easily-mined deposits of Kentucky – where environmental legislation is also less onerous than in other states – and a revival of the depressed Pennsylvania anthracite industry, based on the fuel's high energy content and clean burning properties.[6]

Eastern Interior   This Basin is centred on Illinois, but extends into Western Kentucky and Indiana.[7] It has supplied 20–25 per cent of national production for many years and contains a slightly lower proportion of US coal reserves. Deposits are mainly medium-grade bituminous, with low ash but a high sulphur content usually above 3 per cent. Over half of present production is from generally large surface mines, and this should remain the case in the future, although increasing difficulties will be experienced with thin and deeply buried seams, and opposition to the disturbance of prime agricultural land. For these reasons, underground mines will grow in relative importance by the year 2000. Most output goes to utilities all over the Mid-West and South, with the largest markets in Indiana, Illinois, Kentucky, Missouri, Tennessee, Wisconsin, Alabama, Florida and Georgia. Some 3mt per annum is also sold to coking plants. Forecasts suggest that production from the Basin will expand by three or four times by the year 2000, as a result of the large adjacent industrial and utility markets, good transportation facilities and a private, distributed pattern of land ownership facilitating the rapid expansion of new mines.

Western Interior   These deposits of generally low-quality bituminous coal extend for almost 190 000 square kilometres across the Mid-Western states of Missouri, Iowa, Kansas, Oklahoma, Arkansas and North Texas. Exploitation has been hampered by the depth and thinness of seams, and their often high sulphur content, causing the closure of many mines. Those which survive produce surface-mined coal for limited local markets, mainly in Missouri, although some 2mt per annum of metallurgical grades are mined on the Arkansas/Oklahoma border. Production is likely to remain low in the short–medium term, but in the long term, large resources may be exploited by new extraction methods such as *in situ* gasification.

Western coalfields account for half of US reserves, and a higher percentage of resources, although production amounted to only a third of national output in 1981.[8]

The majority of reserves are found in the Northern Great Plains region, particularly the Powder River and Fort Union basins of Montana, Wyoming and North Dakota. These are mainly subbituminous and lignite, with low moisture and sulphur contents. Seams are generally thick, undisturbed and near to the surface – one Wyoming bed, the Wyodak, is the largest unbroken concentration of coal in the USA, stretching for 200 kilometres with a thickness of 8–40 metres, and containing 15bt of surface-mineable coal (and a further 85bt to 2 000 foot depth). Exploitation is invariably by surface mining and sufficient reserves exist to allow huge increases in production by this method in coming decades.

The Rocky Mountain region has been greatly affected by tectonic activity and exhibits the greatest variety of rank and geological setting of any US coal province. Many deposits are disturbed, or buried at great depths, and are only likely to be won by the development of new mining techiques. Most interest, and exploration, has been centred on the low-sulphur bituminous and sub-bituminous coals of the Green River Basin, Wyoming; the Unita and San Juan River basins of Colorado, Utah and New Mexico; and the Black Mesa field of Arizona. These are mostly won by surface mining, but the high-quality coals of Colorado and Utah are largely mined by underground methods.

As a result of low production costs, and premiums for low-sulphur coal, Western coal output expanded dramatically in the 1960s and 1970s, and reached 268mt in 1981. These factors will continue to operate in the future, and the bulk of the incremental growth in US coal consumption is projected to be met from Western sources (see Table 4.1). At the same time, various constraints – which will not impede other coal provinces to the same extent – will continue to inhibit expansion or to divert it to certain areas.[9] These include concentration of land ownership (some 60 per cent of Western lands are owned by the Federal government, and further large areas are contained in Indian reser-vations); resource shortages, particularly of water; ecological fragility of many coal-containing areas; pressures for alternate uses and high transport costs to distant markets. The latter factor will encourage an existing tendency for coal-using industries to move to producing areas but, if rates are too high, this will be more than offset by lost sales.[10] Thus, one study by the Exxon oil corporation has estimated that half of future US coal-derived synfuel production will be in the Powder River region of Wyoming, whilst several projects for the gasification of North Dakota and Montana coals seem likely to be in operation by the late 1980s.[11]

New developments are also expected to concentrate on deposits in and around the Utah Great basin, due to the availability of water, the high quality of coal, and its accessibility to West Coast ports.

A survey by the Department of Energy has predicted that Western coal output will rise from 234mst in 1980 to nearly 600mst per annum in 1990, when it will represent 40 per cent of national production. Of this, 287mst should come from forty-seven Wyoming mines, 93.7mst from fifteen mines in Montana, 58.7mst from thirteen North Dakota mines, 39.3mst from fifteen mines in New Mexico, 39.2mst from thirty-four Colorado mines (twenty-one of them underground) and 37.8mst from twenty-eight underground mines in Utah.[12]

Other coal deposits are found in the Pacific and Gulf coastal states, which have been developed for local markets. This tendency will continue, especially with regard to the lignite deposits of Texas and

TABLE 4.1   USA coal production by region and mining method: history and projections for three base scenarios, 1965–95 (million short tons)

| Region | Mining Method | History[a] | | | Projections | | | | | | | | |
| | | 1965 | 1973 | 1978 | 1985 | | | 1990 | | | 1995 | | |
					Low	Mld	High	Low	Mld	High	Low	Mld	High
East	Surface	168	235	269	266	266	266	138	139	139	78	78	78
	Deep[b]	324	289	227	489	488	487	680	688	698	816	835	838
	Total[b]	492	524	496	755	754	753	818	826	837	894	913	917
West	Surface	11	57	154	348	347	348	440	468	467	624	726	725
	Deep	9	10	15	27	28	29	46	49	49	74	77	76
	Total[b]	20	68	169	375	376	376	487	517	516	698	802	801
National	Surface	189	298	427	614	613	614	579	607	606	702	803	803
	Deep	338	300	243	516	517	515	726	737	747	890	912	915
	Total[c]	527	599	670	1129	1130	1129	1305	1343	1353	1592	1715	1718
Total Production (quadrillion Btu).		13.4	14.4	15.0	24.9	25.0	24.9	28.5	29.3	29.5	34.3	36.7	36.8

[a]   Source of historical data is Volume 2 of the EIA *Annual Report to Congress*, 1979.
[b]   Coal production includes only bituminous and lignite coal.
[c]   The historical national total includes anthracite coal production, which is mined in the East, mostly in Pennsylvania.

NOTE   Data may not add to total shown due to independent rounding.

SOURCE   US Energy Information Administration, 'Annual Report to Congress 1979, Volume 3', (1980).

Louisiana. Enormous coal resources of 130bt are estimated to exist in Alaska, mainly in the Northern Arctic zone. Production 'from the Nenona field, near Fairbanks, amounts to 1mpta, and may be expanded to meet local markets. Coal deposits around Anchorage may also be developed for export purposes,but Alaska is unlikely to become a major producer until well into the next century, if at all.

ORGANISATION

Until the 1970s, coal was the least regulated of US energy industries, but a host of government measures on mine health and safety, environmental protection and so on, has changed the situation.[13] Recently, coal policy has been formulated by a bewildering variety of agencies, whose lack of co-ordination and technical expertise has been widely criticised.[14] However, Republican victories in the 1980 Presidential and Congressional elections has led to some reduction of government involvement.

The legislative framework of coal policy is provided by Congress, both elements of which – Senate and House of Representatives – contain organised coal caucuses. The main executive responsibility lies with the Federal government, particularly the Departments of Energy and Interior. Energy Department duties – which were trimmed by President Reagan – include strategic energy planning, such as the formulation of National Energy Plans and regulation of some aspects of energy production and use. In the case of coal – administered through its Fossil Energy Division – this includes establishing long-term production goals for federally-owned resources; controlling some aspects of coal leasing; regulating the types of energy used in large user facilities under the 1978 Power Plant and Industrial Fuel Use Act; encouraging research and development of new coal technologies and regulatory price setting.

The Department of the Interior has responsibility for most aspects of coal production, through such divisions as the US Geological Survey, Bureau of Mines, Office of Surface Mining, Reclamation and Enforcement and Bureau of Land Management. Of particular importance is its role under the 1976 Federal Coal Leasing Amendment Act, exercised largely through its Office of Coal Leasing.[15] Following criticisms that leases on Federal coal-bearing lands were being let too cheaply, and for too long periods, few new leases were granted during the 1970s, despite the fact that Federal lands contain a large proportion of national coal reserves. Under the 1979 Federal Coal Management Programme, leasing is now resumed, with higher royalties, and guarantees that they will be 'diligently' developed within ten years. Factors in determining the leases to be offered include contributions to the vitality of the coal industry, maintenance of reasonable prices to consumers and the minimisation of environmental damage, based on twenty 'unsuitability' criteria. Although large amounts of reserves are expected to be sterilised as a result of this Programme, the Interior Department has estimated that production from Federal lands will rise from 100mst in 1980 to 220mst per annum by 1985.[16]

In addition to some functions exercised by other government departments, the coal industry is also regulated by several quasi-independent Federal agencies. The most controversial of these is the Environmental Protection Agency (EPA), which exercises powers under the 1969 National Environmental Policy Act, various Clean Air Act Amendments, the 1977 Clean Water Act and the 1976 Resource Conservation and Recovery Act. Also of great significance is the Occupational Health and Safety Administration with jurisdiction over many aspects of mining operations.

Under much of the legislation cited above, implementation is shared between Federal and state governments, which also possess extensive powers in their own right. One controversial use of these has been the levying of taxes on coal production, especially in Wyoming and Montana. Similar measures have been taken by the Indian reservations, which contain an estimated 30 per cent of US coal reserves and have a great deal of autonomy in controlling their exploitation. Since the mid-1970s, attempts have been made to encourage production and use (by minemouth electricity generation) by Indian-controlled enterprises, and these may become a powerful economic force by the end of the century.[17]

Coal producers in the USA are represented by the National Coal Association. Companies organised by the United Mineworkers (UMW) are also members of a seperate negotiating body, the Bituminous Coal Operators Association (BCOA). User interests are mainly represented through the Edison Institute and the Electrical Power Research Institute. Both producers and users have complained bitterly of the constraints and added expense caused by government regulation, with one company, Consolidation Coal, claiming that these amounted to 50 per cent of underground, and 35 per cent of surface, mining costs.[18] However, against this must be set the fact that external social costs have been reduced (for example, damage caused by pollution), as well as the substantial amounts of government aid–totalling $6.8 billion to 1976 – which the industry has received.[19]

It is also true that the profitability of the coal industry was transformed during the 1970s. From 1969, coal prices began to rise, with the result that, between 1972 and 1976, the value of national coal output rose by 187 per cent, from $4.6 to $13.2 billion – against a mere 14 per cent increase in volume.[20] As a result, coal-company returns on net worth rose from an average of less than 11 per cent in the 1960s to 21 per cent in 1976. Although the situation changed slightly during the late 1970s, the industry ended the decade in far better financial shape than it began it.

In parallel with this change has been a trend towards concentration, with a reduced number of companies operating fewer and larger mines. Between 1955 and 1976 the top fifty operating companies increased their share of national production from 54.7 per cent to 64.5 per cent and suffered less severely from fluctuations in demand than smaller companies.[21] This trend was encouraged by their ability to achieve assured markets guaranteeing future sales to large, dependent consumers, who place great emphasis on security of supply. In 1974, 85 per cent of all coal output was either sold under long-term contract, or produced by captive mines, and the percentage will have increased rather than diminished since that date.[22] The existence of these relationships also aids the financing of new investment, whose costs can be substantial – in 1977, for example, the average capital needs of an underground mine were $65 per ton of production, compared to $96 for an Eastern surface, and $11 for a Western surface mine (see Table 4.2).[23] Finally, larger companies have generally proved better able to bear the costs of health, safety and environmental measures.

Concentration has been accompanied by an influx of non-coal interests. Utilities have joined steel producers as operators of captive mines, whilst oil and mining corporations have taken advantage of the generally low prices of coal-company stocks and long-term coal leases to build a strong position.[24] In 1980 only one of the ten leading producers (business of controlling company in brackets) – Peabody (consortia of mining and other interests), Consolidation (oil – Conoco), AMAX (mining), Texas Utilities (utility), Island Creek (oil – Occidental), Pittston (independent), NERCO (utility), Arch (oil – Ashland), US Steel (steel), AEP (utility) – was an independent company, and this category owned under 10 per cent of privately owned coal reserves.[25] By contrast, oil-company subsidiaries accounted for 22 per cent of national coal output and controlled 41 per cent of privately-owned coal reserves

TABLE 4.2 Typical underground and surface mine costs in the USA[a] (1977 dollars, short tons)

(*Annual tonnage*)	*Underground*[b] (*500 000 tpy*	*Surface*[c]		
		Eastern (*100 000 tpy*)	*Midwest* (*500 000 tpy*)	*West* (*5 000 000 tpy*)
Initial capital investment[d]	$22 966 500	$6 892 230	$17 501 000	$30 221 500
Deferred capital investment	$9 697 500	$2 667 300	$10 087 800	$22 425 500
Total investment[e]	$32 664 000	$9 559 930	$27 588 800	$52 647 000
Capital investment per ton of production	$65.33	$95.59	$51.18	$10.53
Operating per year cost[f] per ton	$8 201 700 $16.40	$1 515 598 $15.16	$4 492 753 $8.98	$11 438 944 $2.29
Selling price per ton, 8 % DCF[g]	$18.14	$19.51	$11.72	$2.70

SOURCE US Energy Information Administration, Economic Analysis of Coal Mining Costs for Underground and Strip Operation (1978).

[a]   Excluding loading and cleaning facilities.
[b]   Representative of room-and-pillar, continuous miner drift operation in a 72-inch coal seam located 700 to 800 feet below surface.
[c]   Representative of a typical Eastern contour mine, Midwestern area mine, and Western strip mine.
[d]   Costs for materials and equipment are based on July 1977 prices.
[e]   Assuming a 20-year mine life.
[f]   Includes operating, maintenance and supervisory personnel, power, operating supplies, and overhead items; wages and union welfare payments are considered as of 6 December 1976 under the Bituminous Wage Agreement of 1974.
[g]   Selling prices are based on an 8 per cent discounted cash flow (DCF) which takes into account all cash flows over the entire life of the operation and adjusts them to start-up time for full production, using compound interest procedures.

in 1978 – since when further expansion has taken place. It is predicted that by 1985 they will account for 45 per cent of production and the figure will probably reach 50 per cent by the 1990s.[26]

The role of the oil companies has caused considerable controversy, centred on the possibility of 'energy monopolies' restricting competition and raising prices. In defence, the oil, and other non-coal, companies claim that their capital, and managerial and marketing expertise, is necessary if the industry is to meet its ambitious future targets. Several government investigations have also found that coal markets are generally 'workably competitive' and should remain so if Federal leasing policy avoids further concentration in Western coal production.[27] Some consumers have remained unconvinced, notably the Tennessee Valley Authority (TVA) which has adopted a policy of supporting independent coal suppliers in order 'not to be at the mercy of the Severn Sisters and their fellow travellers'.[28]

The era of large coal producers has been associated with the rise of surface mining, which accounts for two-thirds of US output. In the West and Interior this has usually taken the form of area and open-pit mining, whilst in Appalachia contour and auger methods have been preferred. Productivity has generally improved more, or declined less, rapidly than in underground mines, largely due to increases in the unit areas mined and the increased size and efficiency of new equipment. Its economics have also benefited from high coal recovery rates and the fact that surface-mining skills are similar to those in other machine-operating industries, thus reducing training costs. However, equipment appears to have reached its maximum size which, together with the effects of increasing strip ratios (overburden to seam thickness) in many areas, will inhibit future productivity increases.

As accessible strip-mineable deposits begin to be worked out by the 1990s, the importance of underground mining can be expected to grow again (see Table 4.1). Until now, the relatively shallow depths of US mines have favoured room-and-pillar methods, using continuous (64 per cent of output) and conventional (28 per cent of output) mining techniques. Although improved productivity and safety can be expected from the development of both these methods, the higher recovery rate allowed by longwall mining is expected to favour increased use of this technique, which provides under 10 per cent of total output. This will further increase the industry's future capital requirements, which will already be swollen during the 1990s by the move to underground production. By 2010, investments of $139–305 billion (in 1976 prices) will be necessary, according to one survey.[29] A further burden will be imposed by the likely trend towards coal preparation as a means of improving quality and reducing sulphur content, although this may be offset by savings in handling and pollution-control costs at user facilities.

In 1980, the coal industry had 245 000 employees, of whom 75 per cent worked underground. A doubling or trebling of output, as has been projected, will require a considerable increase in the labour force, if productivity levels remain constant. In fact, between the large-scale introduction of mechanisation before the Second World War and 1969, productivity – measured as output per manhour – increased rapidly, and grew at 5.8 per cent per annum in the 1960s. After 1969 it fell, especially in underground mines, at an average rate of 3.8 per cent per annum, and by 1977 productivity per worker was only 72 per cent of its 1968 level. Although the question is controversial, the important causes of this decline appear to have been the requirements of the 1969 Coal Mine Health and Safety Act and environmental legislation; re-opening of high-cost capacity as a result of coal price increases in the 1970s; lack of skilled manpower and consequent hiring of large numbers of young, untrained workers; less advantageous mining conditions; the impact of strikes and absenteeism; shortages of capital and the diversion of resources to exploration and other immediately unproductive expenditure.[30] Against this must be set marked improvements in mine safety, with the number of fatalities per million work hours declining from 1.0 in 1970 to 0.35 in 1978. A sharp increase in productivity– in the order of 8–10 per cent – occurred in 1980, and the signs are that further increases might be possible in the 1980s.

Recent decades have also brought marked changes in the characteristics of mineworkers. In 1960, production workers constituted 88 per cent of the labour force, but by 1977 this had fallen to 82 per cent. Most had become better educated and trained, whilst their average age declined. In 1967, only 20 per cent were under 34 years of age, compared to over 50 per cent in 1978, a fact which, despite initial problems such as training, should provide a secure and stable labour force for the future.[31] At the end of the 1970s, the hiring of women also increased, and several companies have agreed training quotas with an eventual goal of a 33 per cent female workforce. Finally, wage rates have improved and are now amongst the highest of any industrial workers. As a consequence, the social and economic position of mining communities have also changed for the better, although problems of education, housing, health care and soon, still occur in many areas of Appalachia and Western 'boomtown' communities.[32] These changes have been reflected within the United Mineworkers of America (UMW), the main coal union, whose centralised autocratic structure – forged under the tough leadership of John Lewis – brought great success in its past negotiations with employers.[33] However, with the job losses and low wage settlements resulting from the industry's depression in the 1960s, tension increased between the membership and an ageing union bureaucracy, culminating in 1969 when Joe Yablonski, a reform candidate for the UMW Presidency, was murdered, with the involvement of some officials. Although reform groups eventually gained control and made the union more effective, its position has been weakened by the growth of non-union mines, particularly in the West, and the existence of rival unions, principally the International Union of Operating Engineers, the Progressive Mineworkers of America and the Southern Labour Union. Its problems were also reflected in numerous disputes and 'wildcat' unofficial strikes at local level. From 1970 to 1976, there was an annual average of 947 stoppages in the bituminous coal industry, compared to an annual average of 184 in 1960–9.

The 1978 strike by the UMW was a decisive turning point. Some companies hoped to break the union's power, but its eventual, partial, victory ensured its survival. Afterwards new grievance procedures, changes of attitude by both employers and union officials, and the changing economic position of the industry created an improved climate and strikes were reduced. Although the 1981 strike turned the clock back to a certain extent, it still appears that coal labour relations may be more successful in the 1980s than they have been in the recent past.

MARKETS

Until the Second World War, coal in the USA was a widely-used, basic raw material; since then, its markets have gradually been restricted to power generation, steel and exports. Although a decline in relative transport costs – from 60 per cent of the delivered price of coal in the 1930s to 32 per cent in 1976 – has helped create a national market for coal, most sales are regionally based. Typically, four such regional markets are identified – Appalachia, Midwest, Northern Great Plains and South West – each clustered around a main producing basin.[35] (see Table 4.3).

Although coal has become relatively less important as a boiler fuel during this period, the rapid growth in electricity demand increased the consumption of coal for this purpose from 72mst in 1945 to 593mst in 1982. Most of this was burnt in base-load plant of over 300MWe capacity, concentrated in the Ohio Valley states. Greater expansion was prevented by the perceived economic and other advantages of oil, gas and nuclear power for base-load purposes. During the 1970s, the rising price of these alternatives, together with the insecurity of oil supplies and

TABLE 4.3 USA coal prices by region and coal type: history and projections for three base scenarios, 1965–95 (1979 dollars per million Btu, FOB mine)

Region	Coal type by rank and Sulphur Level[b]	History[a]			Projections								
		1965	1973	1978	1985			1990			1995		
					Low	Mid	High	Low	Mid	High	Low	Mid	High
Northern Appalachia	Bituminous/HS	—	—	—	1.33	1.33	1.33	1.47	1.47	1.50	1.61	1.61	1.62
Central Appalachia	Bituminous/LS	—	—	—	1.77	1.77	1.77	1.93	1.93	1.93	2.12	2.12	2.10
Midwest	Bituminous/HS	—	—	—	1.21	1.21	1.21	1.34	1.34	1.34	1.43	1.43	1.43
Great Plains	Subbituminous/LS	—	—	—	0.51	0.51	0.51	0.51	0.51	0.51	0.51	0.51	0.52
Rockies	Bituminous/LS	—	—	—	1.10	1.12	1.13	1.24	1.25	1.25	1.35	1.37	1.36
National Average	All types	0.39	0.56	1.06	1.26	1.26	1.26	1.38	1.36	1.37	1.42	1.38	1.38

[a] Source for derived historical data is Volume 2 of the EIA *Annual Report to Congress*, 1979.
[b] LS denotes 0–0.67 pounds of sulphur per million Btu. HS denotes greater than 1.68 pounds per million Btu.

NOTES —indicates not available.
Coal prices for bituminous and lignite coal only.

SOURCE US Energy Information Administration, 'Annual Report to Congress Volume 3' (1980).

political opposition to nuclear power, have encouraged greater use of coal, although in the short term this has been offset by low rates of electricity demand growth (3.5 per cent per annum since 1973, compared to 7–8 per cent in the 1960s) and consequent generating over-capacity. These factors, and government intervention – including the banning of oil and gas combustion at some sites, and subsidies for conversion to coal – mean that coal will be the main source of incremental growth in generating capacity during the 1980s and beyond. According to the WOCOL Report, installed coal-fired capacity should rise from 200GWe in 1977 to 284–305GWe in 1985, with much of the increase occurring in the West.[36]

Estimates of utility coal consumption (see Table 4.4) in the year 2000 range from 800 (WOCOL Low Coal, IEA) to 1 170mtce per annum (WOCOL High Coal) with official projections suggesting a market for 1.06–1.14bt per annum by 1995.[37] However, delays in power-plant construction demand and costs of pollution control may render such forecasts optimistic. Growth in coal consumption may also be limited by increased efficiency of use, including new generating technologies; growth of combined heat and power plants; more flexible electricity systems making greater use of regional or national grids and high transport costs. In the latter case, one study has found that utility use of Western coal is more sensitive to transport than mining costs.[38] Domestic sales of coking coal during the 1970s were hit by recession in the steel industry, coke imports and the development of new steel-making methods, and stood at 40mst in 1982. The steel industry accounted for 90 per cent of these sales, of which 60 per cent was produced in their captive mines. Lack of investment has led to problems in coke-making, with capacity falling from 58.4mst per annum in 1970 to 50mst per annum in 1979.[39] In 1978, 69 per cent of this capacity was over twenty years old, and 40 per cent over twenty-five years old, compared to an average oven life of thirty years. As a result, 5mt of coke was imported in 1979, mainly from West Germany, and imports are likely to remain substantial during the early 1980s. After that point, government-supported investment programmes should create new oven capacity and a consequently increased market for domestic coking coal. Forecasts for the year 2000 suggest that this will range between 92 (IEA) and 110 (WOCOL High Coal)mtce per annum, while official projections say 79–80mt per annum by 1995.

Purchases of coal by industrial and residential/commercial consumers rose slightly in the 1970s, reaching 75mst in 1982. This is used to fire boilers, provide process heat and as a raw material, with the bulk of

TABLE 4.4 USA coal consumption by end-use sector: history and projections for three base scenarios, 1965–95 (million short tons per year)

| Sector | History[a] | | | Projections | | | | | | | | |
| | | | | 1985 | | | 1990 | | | 1995 | | |
	1965	1973	1976	Low	Mid	High	Low	Mid	High	Low	Mid	High
Electric utility	245	389	481	740	737	739	861	884	890	1055	1115	1136
Industrial[b]	132	79	73	219	223	221	241	248	249	268	280	258
Synthetics	—	—	—	12	12	12	19	27	31	50	101	105
Domestic coking	95	94	71	74	74	73	79	78	77	80	79	79
Total Domestic Consumption	472	563	625	1045	1046	1045	1199	1237	1247	1452	1575	1577
Net Exports	52	54	38	85	85	85	108	108	108	143	143	143
Change in Stocks[c]	3	(18)	7	NA	NA	NA	NA	NA	NA	NA	NA	NA
Total production	527	599	670	1129	1130	1129	1305	1343	1353	1592	1715	1718
Total Btu (quadrillion)	13.4	14.4	15.0	24.9	25.0	24.9	28.5	29.3	29.5	34.3	36.7	36.8

[a] Source of historical data is Volume 2 of the EIA *Annual Report to Congress*, 1979.
[b] Includes small amounts of coal used by the residential, commercial, and transportation sectors.
[c] Includes changes in stocks, imports losses and unaccounted for.

NOTES NA indicates not applicable.
        —indicates not available.
        Coal consumption for bituminous, lignite, and anthracite coal.

SOURCE US Energy Information Administration, 'Annual Report to Congress Volume 3' (1980).

consumption occurring in the chemical, paper, cement, metal products and food industries. Unlike other sectors, a high percentage of requirements are purchased on spot markets, although more long-term contracts are expected to be signed in future. Most forecasts suggest that this market will grow substantially, aided by developments in gasification technologies, with demand forecasts for 2000 ranging from 125 (WOCOL Low Coal) to 220 (WOCOL High Coal) and a staggering 458mtce per annum (EMF).[40] Official forecasts project sales of 258–280mt per annum by 1995.

The sector may be able to grow more rapidly than others because of the relatively small-scale, decentralised nature of its plant, with less intense environmental or other opposition, and the development of new user industries, especially petroleum production in shale-oil plants or steam extraction from heavy oil reservoirs. However, the costs of conversion – one study has estimated that coal-fired boilers can cost from two to five times as much as oil- or gas-fired equivalents – and resistance to change may be serious impediments to progress.[41]

American interest in coal conversion intensified in the 1970s, and a large number of demonstration liquefaction and gasification plants are now operating or under construction. A major boost was provided by the 1980 Synfuels Act, under which $20 billion was made available to a Synfuels Corporation, with the possibility of a further $68 billion being made available by 1990. The aim of these measures was to reach a synfuel output of 2mbdoe by 1992, much of which would be derived from coal conversion. However, funding and targets were reduced by the Reagan administration and the corporation now aims to develop technology for future rather than present application. Estimates of the demand for coal feedstock for this purpose in 2000 have ranged from 50mtce per annum (WOCOL Low) to Exxon predictions of an annual synfuel output of 4mbdoe, requiring 645mst per annum of coal.[42] Projections for 1995 suggest a market of 50–105mst per annum.

Official intervention in this field was cut back after the 1980 Presidential and Congressional elections, and several studies have suggested that high capital costs and technological difficulties will postpone the large-scale construction of liquefaction plants until the late 1980s, and SNG facilities until the 1990s. Economic factors may also favour the development of other unconventional energy sources, such as heavy oil deposits, in preference to coal-derived products. For these reasons, one study has predicted that output of all synfuels will only reach 0.7–1.2mbdoe by 2000.[43]

In 1982 exports were 106mst (96mt), well down from 1981. The

WOCOL Report predicted export levels of 200–400mtce per annum – considerably higher than most previous estimates – by 2000. Subsequent government studies endorsed the lower end of this range as a feasible policy target, although there is scepticism in other quarters as to whether the many problems can be overcome.[44] These doubts are reflected in the preference of coal importers for non-US coal, where this is possible.[45]

In 1981 the USA also had imports of 2mt, mainly of South African steam coal to Gulf Coast and South-Eastern consumers. High internal freight charges and/or supply bottlenecks could produce an increase in imports to these markets.

ENVIRONMENT

Environmental factors were a major constraint on increased coal production and use during the 1970s, and this situation seems likely to continue in the future. With coal production, the effects of surface mining have caused the greatest controversy. In 1977, coal was responsible for most of the 5.7 million acres which had been strip-mined in the USA, and of the 1.66 million acres which were judged to be in need of reclamation work. Studies have suggested that areas affected by surface mines – which can spread beyond the actual mining site – would reach 20 million acres by the year 2000.[46] To date, attempts at reclamation, especially in the arid and ecologically-fragile West – where costs were estimated at $3 700–6 673 per acre in 1979 – have had mixed success, and one report has bluntly stated that some coal lands must be regarded as 'national sacrifice areas' which will never be fully reclaimed.[47]

Control is presently exercised by the 1977 Surface Mining Control and Restoration Act, which also covers the surface facilities of underground mines. Its provisions state that all mining must be licensed by Federal or state authorities; be accompanied by performance bonds; replace topsoil; in the case of steep-slope mining, place no spoil on downslopes; have only limited impact on water quality and quantity; return land to prior use or better, with approximate contours restored, and pay a production levy to restore abandoned lands. In addition, certain areas – especially prime farmland and alluvial valley floors – are specially protected or precluded from mining. The Office of Surface Mining has been charged with inflexibility in operating the Act, and some of its powers have been trimmed or transferred to state govern-ments as a result of court judgements and the 1980 elections.

Other Federal legislation, which applies to both surface and under-ground mining, includes the Clean Air Act, the Resource Recovery and Conservation Act and the Clean Water Act, this latter being of particular importance in the arid West (see Fig. 4.2). Underground mining also causes major subsidence problems, causing damage with an annual cost of $30 million.[48]

Coal use was more affected by the Clean Air Act, and its amendments, than any other factor in the 1970s.[49] Under its terms, the Environmental Protection Agency (EPA) has identified sulphur dioxide, nitrous oxides, carbon monoxide, photochemical oxidants, hydrocarbons and par-ticulates as criteria pollutants, and formulated National Ambient Air Quality Standards (NAAQS) to stipulate the maximum permissible limits. These must be enforced by state governments under 'State Implementation Plans' (SIPs). Since 1971 all large new coal-using facilities, as well as some mines, have been subject to New Source Performance Standards (NSPS), whose main provision has been an upper limit of 1.2 lb sulphur dioxide, 1 lb particulate and 0.9 lb nitrous oxide emissions per million Btu (MBtu). For much of the 1970s this encouraged a consumer preference for low-sulphur, 'compliance' coal from the West and Kentucky, as opposed to the high-sulphur Appalachian and Interior coals which required expensive flue-gas scrubbing, and other pollution controls, to meet this standard. As a result of 1977 Congressional amendments, designed to redress the balance between East and West, the EPA introduced new standards in 1979, applying to all major coal-using facilities built since September 1978. These maintain the upper limit, but classify coals into three categories, each subject to a different level of sulphur dioxide reduction. Category 1 coals, producing 2 lb or less per MBtu, must reduce emissions by 70 per cent. Category 2 coals (2–6 lb per MBtu) must cut them by 70–90 per cent, whilst Category 3 coals, with 6–12 lb per MBtu, must reduce emissions by 90 per cent. In most cases these requirements will necessitate the installation of flue-gas scrubbing systems, although some flexibility can be achieved by better coal preparation, to wash away inorganic sulphur, and other techniques. In addition, separate regu-lations have reduced permitted NSPS particulate levels to 0.03 lb per MBtu and nitrous oxide emissions to 0.5–0.8 lb per MBtu, depending on the coal used.

Air-quality factors also influence the siting of coal-producing and –

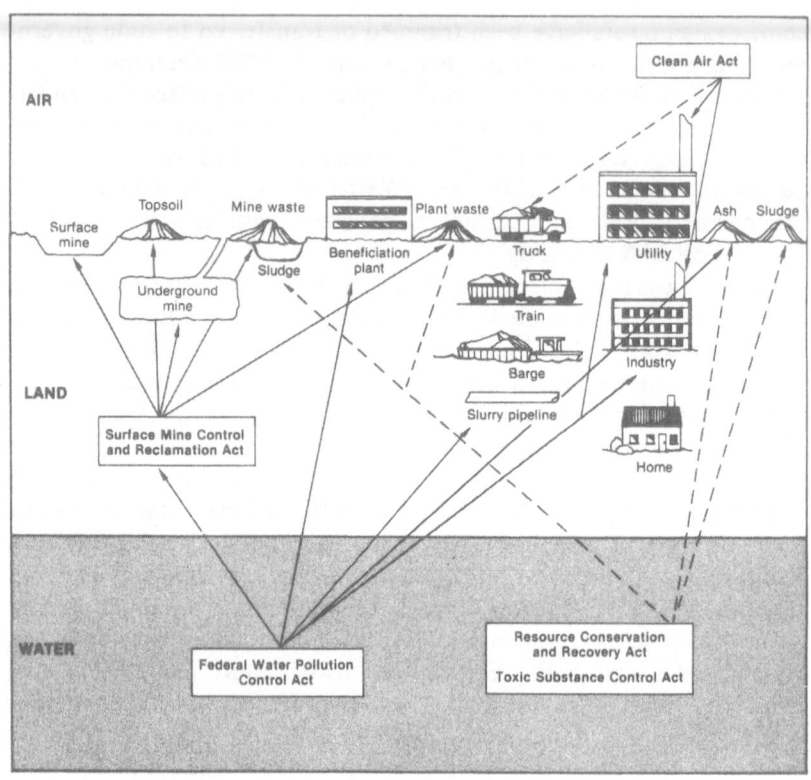

FIGURE 4.2   Jurisdiction of US Federal legislation

SOURCE Congress of the United States, Office of Technology Assessment, 'Direct use of coal' (1979.

using facilities, especially in areas which have exceptionally clean air – and are subject to Prevention of Significant Deterioration (PSD) restrictions – and 'Non-Attainment Areas', where NAAQS standards have not been met. The sitting and operation of user facilities are also affected by numerous other Acts, many of which are enforced by the EPA. These include the Resource Recovery and Conservation Act – under which standards for powerplant wastes are introduced – Clean Water Act, Endangered Species Act, Toxic Substances Control Act, and Coastal Zone Management Act. Such measures have been a major factor in increasing power-plant construction times from 3–4 years in 1970 to up to 10 years in the 1980s.

Much controversy has surrounded the question of whether environmental measures have restricted, and will continue to restrict, coal production and use. According to industry, the EPA standards are unrealistically high and cannot be met by existing technologies – or when they can, are prohibitively expensive. Both claims are disputed by the EPA, whose studies claim that new, more efficient control technologies will be developed, and that coal remains competitive with other fuels even when the costs of environmental measures are taken into account.[50] However, after 1979, and especially since the 1980 elections, regulations became less stringent. In particular, emission controls are now applied to the aggregate emissions of plants – the 'bubble' concept – rather than individual units.

In the 1980s the main issues of the 1970s – sulphur-dioxide emissions and acid rain, carbon dioxide and surface-mining reclamation – are likely to be joined by the question of hazardous wastes from both flue-gas scrubbing and coal conversion, the effects of large-scale coal transport and water quality, which several reports have singled out as a particularly critical factor. Already twenty-one out of 106 major hydrological basins in the USA have severe water shortages, and a further sixteen are predicted to join them by 1985 – with the problem being particularly acute in the West and the Ohio River Basin. It is difficult to avoid the conclusion that 'expansion in coal regions will be restricted physically by lack of resources, such as water, and institutionally, by environmental protection laws and regulations'.[51]

However, although environmental factors may preclude coal mining and use in certain areas, surveys have shown that the costs of protection measures in areas where these activities are allowed are unlikely to significantly constrain consumption in the future.[52] In addition, in some areas the problems may be eased by a partial consensus between the industry and the environmentalists, which have been brought together in the National Coal Policy Project. Both sides have agreed that: coal can be mined in many areas without environmental damage if proper precautions are taken; there is a need for marginal cost-energy pricing; coal-burning facilities should be sited in user areas; the EPA should allow some exemption from existing regulations for the trial of new technologies; siting procedures should be streamlined and experiments should be made with pollution taxation and that coal transport should be deregulated.[53]

The 1980 elections also brought to power a party and a president who are less sympathetic to environmental issues than their predecessors. Much environmental legislation has been amended or repealed, and

Western lands have been opened up for the exploitation of coal.
Nevertheless, the American political system provides many opportunities for environmental movements to work towards their goals, and
the subject will remain extremely controversial throughout the 1980s.

INFRASTRUCTURE

The regional pattern of US coal production and consumption has
produced a corresponding pattern of transport. Over the next few
decades this position will alter as coal is shipped over greater distances
to both home and export markets.[54] Many observers have identified
these changing transport requirements, and the substantial resources
needed to meet them – capital costs have been estimated at $67 billion to
the year 2000 – as one of the major constraints on future coal
expansion.[55] Several reports have also identified transport costs as the
major factor in determining the future balance of production between
Eastern and Western America.[56]

In 1980, 64 per cent of coal output was transported by rail, and this
is expected to remain the dominant mode until 2000. This traffic
represented 29 per cent of all US rail freight and was mostly carried on
the Chessie System, Louisville & Nashville, Burlington Northern,
Norfolk & Western, Conrail and Southern lines. Major investment and
improved efficiency will be needed to revitalise US railroads if future
coal-transport projections are to be met. Shortage of rolling stock
limited production in Appalachia during the 1970s, and an estimated
58 000 new, and 113 000 replacement, wagons will be required before
1985.[57] To an extent, this problem will be reduced by increased use of
unit trains – fixed-coupling stock, typically of a hundred 100 ton
hoppers circulating constantly between mine and market – four of
which can deliver some 2mt per annum. Past neglect has also left most
track in poor condition, especially that owned by older railroads in the
North-East, and a third has speed restrictions of 10mph or less. The need
for new track will be accompanied by necessary work on grade (level)
crossings, especially in the West, where roads can be closed for up to 12
hours daily.

To finance this investment, US private railroads will require $13–15
billion of outside capital by 1985, with a need for further public funding
of Conrail and other public enterprises. With government policy set
against subsidy, a large part of future investment must be met from
increased revenue, although controversy has been aroused both within

and outside government as to whether this should be by 'differential pricing', that is charging higher rates for coal than other products, or 'cost-plus pricing' for all commodities. Several court cases have been fought over the question, utilities and others arguing that rail charges are excessively high. However, although these may limit coal use in the immediate future, failure to invest may have an even greater impact in the medium–long term.

Although motor vehicles are involved in 75 per cent of all coal movements, most of this is from and to rail and water terminals, and only 11 per cent of coal output in 1980 was transported solely by this route. This is largely because the low weight and capacity of trucks on public highways, together with high fuel costs, makes their use uneconomic for long-haul transport. Roads in many mining regions, particularly Appalachia, are already inadequate for the volume of coal traffic using them, causing major environmental and safety problems. Rising production and under-investment by public authorities – who need to spend up to four times more on coal roads than others – mean that this situation will continue for the immediate future, and one report has warned that 'Appalachia's coal road problems could well become so severe as to become a bottleneck on coal production'.[58] The eventual solution will be some contribution to highway cost by mining companies and the Federal government, although agreement on the issue seems far away.

Water-transported coal accounted for 11 per cent of national output in 1980, mostly by towed barges on the Mississippi–Ohio and Gulf intracoastal waterways, and Great Lakes bulk carriers.[59] Although many facilities are old, and delays have been experienced in some areas during the 1970s, the energy efficiency of water transport is likely to make it a favoured transport mode, with one study estimating that coal traffic on Mid-American waterways will grow from 131mst in 1976 to 298mst in 2000.[60] Meeting these targets will be helped by the opening of the Tennessee–Tombigbee and other new waterways, but will also require major expenditure to remove bottlenecks, especially limited lock capacity.

Concern at the ability of conventional rail, road and water transport systems to meet future increases in coal traffic has turned attention to slurry pipelines.[61] Such systems comprise a preparation plant at the mine, where coal is crushed and mixed with liquid, a pipeline along which the slurry is pumped, and a separation plant at the end. Most will use water as their slurrying liquid, although methanol and other alternatives have been suggested. Several studies have suggested that such pipelines can be the most economic method of moving high volumes of coal over large distances (see Fig. 4.1) and that large volumes will be transported in this way by 2000. However, major obstacles which face such pipelines include lack of eminent domain (compulsory purchase) rights, difficulty in obtaining water in arid areas and opposition from railroads. However, several states have granted rights of eminent domain and federal legislation to this end seems likely to be introduced in the 1980s.

At present, only one pipeline is in operation, carrying 5.5mst per annum from the Black Mesa mine of Peabody Coal in an 18 inch diameter pipe over 273 miles to a Laughlin, Nevada, power plant. During the 1980s, further lines have been proposed from Wyoming and Colorado to Texas, the Midwest and the West; from mines to power plants within the South-West, and from the Appalachians to Florida. Several of these will probably be constructed, even if rights of eminent domain are not granted. By 2000, it has been estimated that some $9 billion will have been invested in these facilities.[62]

One type of coal use which has no need for its physical transport is minemouth electricity generation, which consumed 11.7 per cent of national production in 1976. Although more coal is likely to be used in this way – or converted into gas or liquids at the mine – problems may be encountered with the transmission network needed to serve distant markets. Concern about both the land requirements and alleged harmful effects of high-voltage cables has caused intense opposition in some parts of the USA, notably Minnesota.

Most US coal exports use ocean shipping, and projected increases will require a minimum of $2.5 billion investment for new port facilities. At present, most coal is sent to East Coast ports, particularly Hampton Roads and also Baltimore and Philadelphia, all of which were subject to some congestion.[63] Small amounts are also exported from Mobile and New Orleans, and these ports are expected to grow substantially to handle Interior and Western coal shipped by pipeline and inland waterway. Some trans-oceanic exports may also be shipped via Great Lakes ports. Demand from East Asia is also expected to lead to several new coal terminals on the Pacific, with San Francisco, Los Angeles, Portland and Seattle as the most likely choices.

COAL AND ENERGY POLICY

Coal expansion has been a central feature of US energy policy since 1973, when the Middle Eastern crisis demonstrated the political and economic consequences of increased dependence on imported oil. Although the practical results of these policies have been disappointing, most forecasts agree that the next two decades will see a major increase in coal production and use, Government projections made in 1980 suggested that output would reach 1.59–1.72 billion tons per annum by 1995 (see Table 4.1) and up to 3.25 billion tons per annum by 2020.[64] Subsequent forecasts increased the 1995 projections to 1.85–1.90 billion tons per annum, of which half would be from the West. From this output, utilities are expected to take 1160–1167 million tons per annum, industrial and residential markets 235–257 million tons per annum, emerging coal technologies (including FBC) 208–298 million tons per annum and coking 74–77 million tons per annum.[65] The WOCOL Report is even more optimistic, forecasting that output may reach 2 bt per annum by 2000, although this has caused some scepticism within the industry itself.

The eventual out-turn will clearly depend upon a variety of factors. Most forecasts assume that economic growth rates are energy/GNP co-efficients will be lower than in the past, reducing the growth in energy usage – to 1.2–2.4 per cent per annum until 2000, according to the WOCOL Report. If the fall is greater than this, coal's entry into new markets will be correspondingly more difficult. Developments in other energy industries will also be of importance, by making them more or less attractive alternatives to reliance on coal. This is particularly true of nuclear power – the current technical, economic and political problems of which are encouraging utilities to install more coal-fired capacity – and oil and gas, where high prices may open up new sources, as in the Western Overthrust belt or by enhanced recovery of existing reserves. The choices which are made will also depend on the movement of coal prices, which one report suggests will rise by 25–50 per cent in real terms to the end of the century.[66] However, the same report has concluded that, in the long run, coal sales will be determined more by the physical opportunities of consumers in a variety of markets to obtain and use coal than by price.

At present coal output is limited by demand rather than supply factors, and substantial unused capacity is available to meet sudden substantial increases in consumption. One report suggests that this situation may persist until 2000, but it is equally conceivable that environmental, infrastructural or financial problems may slow the rate at which new capacity can be opened.[67] Delays in opening up new port and rail facilities could also limit the abilities of coal producers to meet market requirements.

Two considerations of crucial importance to all these questions are the determination – and ability – of the Federal government to make its pro-coal policies successful, and the political acceptability of this within the USA as a whole. During the 1970s, government policy was indecisive and unco-ordinated – as late as 1979 utility and industrial plants were being encouraged to switch from oil to gas rather than coal – and only moderate increases in coal production and use were registered. The decentralised nature of the American political system also gave scope for independent agencies and environmental groups to prevent, or delay, many coal-based projects.

In the short – medium term, the 1980 Presidential and Congressional elections have strengthened the existing trends which favour coal expansion. The easing of regulatory burdens and the opening up of more low-cost coal in Western lands outweighs the loss of Federal assistance in other areas, while continuing oil and gas price rises, and the problems of nuclear energy, are opening up new market opportunities. In the long term, several doubts remain as to whether a massive expansion of coal output can be achieved. As one report has commented, 'mining and burning 2–3 times the present coal output, even if done efficiently and with care, will be difficult (and increasingly expensive) if the contributions of this energy source to air and water pollution and land degradation are to be kept from increasing'.[68] By the 1990s, judgements will also be made about the industry's environmental performance during the expansionary 1980s. The necessary co-ordination of private investment and the multiplicity of US government agencies to provide reliable 'coal chains' between mines and end-users may also prove an impossible task. Finally, the 'energy free market' towards which US energy policy is moving, and which benefits coal in the short run, may rebound against it by encouraging the development of more economically or environmentally advantageous alternatives, or by driving energy prices so high that economic and energy demand growth rates are depressed beyond present expectations, with damaging effects on all energy industries.

5 Red Coal

Most of the world's coal resources, and its present production, are concentrated in countries with Communist governments and Centrally Planned Economies (CPE). Although these are divided into mutually hostile blocs, the nature of their coal industries, and their relative isolation from those of the free market economies – with the partial exception of Poland and, to a lesser extent, China – justify their common treatment.

Most Communist nations are members of the Council for Mutual Economic Assistance (COMECON), which is dominated by the USSR. Its other members are, in Eastern Europe, Bulgaria, Czechoslovakia, the German Democratic Republic (East Germany), Hungary, Poland and Romania and, elsewhere, Cuba, Mongolia, South Yemen and Vietnam. Angola, Laos, North Korea and Yugoslavia also have an involvement in some areas. In 1971 the full members agreed a Comprehensive Economic Co-operation and Integration programme, leading to the specifying of common energy and raw-material objectives in 1976. In terms of coal and most other energy matters, effective action has been restricted to Eastern Europe, where it has been encouraged by the political and economic hegemony of the USSR.

USSR

If the USA is not, and cannot be, the 'Saudi Arabia' of coal, the USSR is the only other contender for the title.[1] It is now the largest coal producer in the world, and possesses almost half of estimated global coal resources – recoverable reserves of 165btce and additional resources of 4432btce. These treasures provide ample security for increases in production beyond the 1981 figure of 719mt.

In fact, although in the 1960s and early 1970s, greater priority was given to oil and gas, the Soviet coal industry never suffered the under-investment and absolute decline of its Western counterparts. Primarily, this was because the energy-intensive industrial growth of this period provided markets for all fuels. A secondary factor was the incentive to export oil and gas to COMECON (for political reasons) and the West (for hard currency), leaving coal to supply a large section of the domestic market. Since 1975, there has been renewed emphasis on coal in Soviet energy planning, and production is scheduled to expand greatly by the end of the century. However, there are several major problems confronting the industry, as became evident in 1979 when, for the first time in two decades, output declined from that of the previous year. Prominent amongst these are the mis-match between coal deposits (mainly in the East) and markets (mainly in the Urals and the West); labour shortages and low productivity; poor organisation and technology, and a deterioration in mining conditions and coal quality in existing fields.

Geography

The geography of the USSR coal industry has changed markedly in recent years and this is likely to continue in the future (see Tables 5.1–5.3 and Fig. 5.1). Today, the main coalfields are:

Donets Basin, the first area in the present USSR to be mined for coal – in 1796 a British traveller, Mrs Guthrie, remarked that she was 'sitting at a cheerful fire made . . . with coals' – this has retained its dominance until the present day, despite the destruction it suffered during the Second World War. In 1978 the basin produced 29 per cent (232mt) of national output, including 50 per cent of all coking coal. Its situation in the Ukraine and South-European Russia makes it ideally placed to supply steam coal to most of the Western USSR, although cheaper coal from the East has been taking some of its markets in recent years. Most of the coking coal produced is consumed within the great metallurgical industries of the Basin.

The age, and increasingly difficult working conditions of the field, make it difficult to maintain existing levels of production, which fell during the late 1970s. Almost 50 per cent of reserves are at depths greater than 700 m, and the average mine depth is reported to be 566 m – double the 1962 figure, and a third deeper than the national average. More than 80 per cent of total production also comes from seams less than 1.2m thick, and a high proportion of longwall faces are working seams less than 0.7m thickness. In addition, methane concentrations can be up to 10 per cent, whilst the quality of coal mined appears to be deteriorating.

TABLE 5.1 USSR: characteristics of major coal deposits[a]

	Type of mining	Explored reserves (billion metric tons)	Thickness of seam (metres)	Depth of mine (metres)	Average calorific value (Btu per pound)	Moisture content %	Ash content %	Share of production in 1978 %
Donetsk	Underground	40	0.9	566	10 900	6.5	19.2	29
Pechora	Underground	8	2.4	454	9 390	8.3	25.1	4
Moscow	Underground	5	2.5	135	4 550	32.3	35.5	4
Kuznetsk	Underground and open pit	60	2.5	262	9 990	10.2	19.0	20
Karaganda	Underground	8	2.5	384	9 250	7.5	28.8	7
Ekibastuz	Open pit	4	10–40		7 250	7.7	39.1	8
Kansk-Achinsk	Open pit	72	NA		6 490	33.0	10.7	4

[a] Source: V. A. Shelest, *Regional'nyye energoekonomicheskiye problemy SSSR*, 1978, pp. 113–16. See also I. I. Novitskiy, *Energeticheskoye toplivo SSSR*, Moscow, 1979, pp. 10–14.

SOURCE US Central Intelligence Agency, 'USSR: Coal Industry Problems and Prospects', (1980).

TABLE 5.2 USSR: production of raw coal, by basin (million metric tons)

	1960	1970	1975	1976	1977	1978[a]	1980[b]	1980[c]	1985[c]
Total	510	624	701	712	722	724	805	725	775
Western USSR	268	301	311	310	306	297	329	289	280
Donetsk	188	217	223	225	223	212	232	208	200
Pechora	18	21	24	26	27	28	31	29	33
Moscow	43	36	34	31	29	29	34	26	25
Other	19	27	30	28	27	28	32	26	22
Urals	59	54	45	46	45	45	39	36	25
Kazakhstan and Central Asia	40	70	102	104	108	115	136	126	165
Karaganda	26	38	46	47	48	48	52	49	55
Ekibastuz	6	24	46	47	50	57	72	68	100
Other	8	8	10	10	10	10	12	9	10
Siberia and Soviet Far East	143	199	243	252	263	267	301	274	305
Kuznetsk	84	113	138	143	145	145	162	150	160
Kansk-Achinsk	10	18	28	29	32	32	42	35	50
Other	49	68	77	81	86	90	97	89	95

[a] Data for 1960–77 from the No. 4 Issue of *Ugol'*, Data for 1978 based on *Ekonomicheskaya gazeta*, No. 5, 1979, pp. 1–2.
[b] Original plan.
[c] Estimated (eventual 1980 total production, 716).

SOURCE US Central Intelligence Agency, 'USSR: Coal Industry Problems and Prospects', (1980).

TABLE 5.3   USSR: cost of coal production, by basin (rubles per metric ton of standard coal equivalent)

	1960[a]	1965[a]	1970[b]	1975[c]	1977
Donetsk (underground)	12.7	14.4	15.4	17.0	17.7[d]
Kuznetsk (open pit)	5.5	6.2	6.6	8.6	9.0[d]
Pechora (underground)	14.3	16.5	15.5	NA	NA
Moscow (underground)	NA	NA	NA	24.1	25.0[e]
Karaganda (underground)	8.2	10.8	10.9	12.8	13.0[e]
Ekibastuz (open pit)	NA	NA	NA	2.5	2.5[d]
Kansk-Achinsk (open pit)	NA	NA	NA	2.4	2.9[d]

[a] *Tsena, sebestoimost' i rentabil'nost' v ugol' noi promyshelennosti*, Nedra, Moscow, 1974, p. 57, and *Territorialnaya differentisiya tsen v tyazheloi promyshlennosti, ekonomika*, Moscow, 1974, p. 21.
[b] *Voprosy ekonomiki*, June 1971, p. 37.
[c] *Planovoye khozyaystvo*, No. 6, 1975, p. 66.
[d] *Elektricheskiye stantsii*, No. 12, 1978, p. 13.
[e] Estimated.

SOURCE   US Central Intelligence Agency, 'USSR: Coal Industry Problems and Prospects', (1980).

The CIA reports that its average calorific value fell by 18 per cent between 1971 and 1979, whilst reports have also suggested that its sulphur content is increasing. These problems – which would prevent much Donets coal from being classified as 'economically and technically recoverable' in the West – result in low levels of mining productivity, which is only half the level of Kuznetsk collieries. This situation is worsened by the high average age of the workforce. In 1978, the Basin required 55 per cent of the USSR mining labour force (450 000 people) to produce only 30 per cent of its output.

Responses to these problems have included the development of shallow deposits on the fringe of the Basin; building large, deep mines in existing mining areas; and new methods of thin-seam mining and methane extraction. However, the CIA estimates that mine depletion is now running at 3.5mt per annum and that planned investments will not be sufficient to replace these losses, causing a decline in production during the 1980s. This pessimism seems to be shared by Soviet planners,

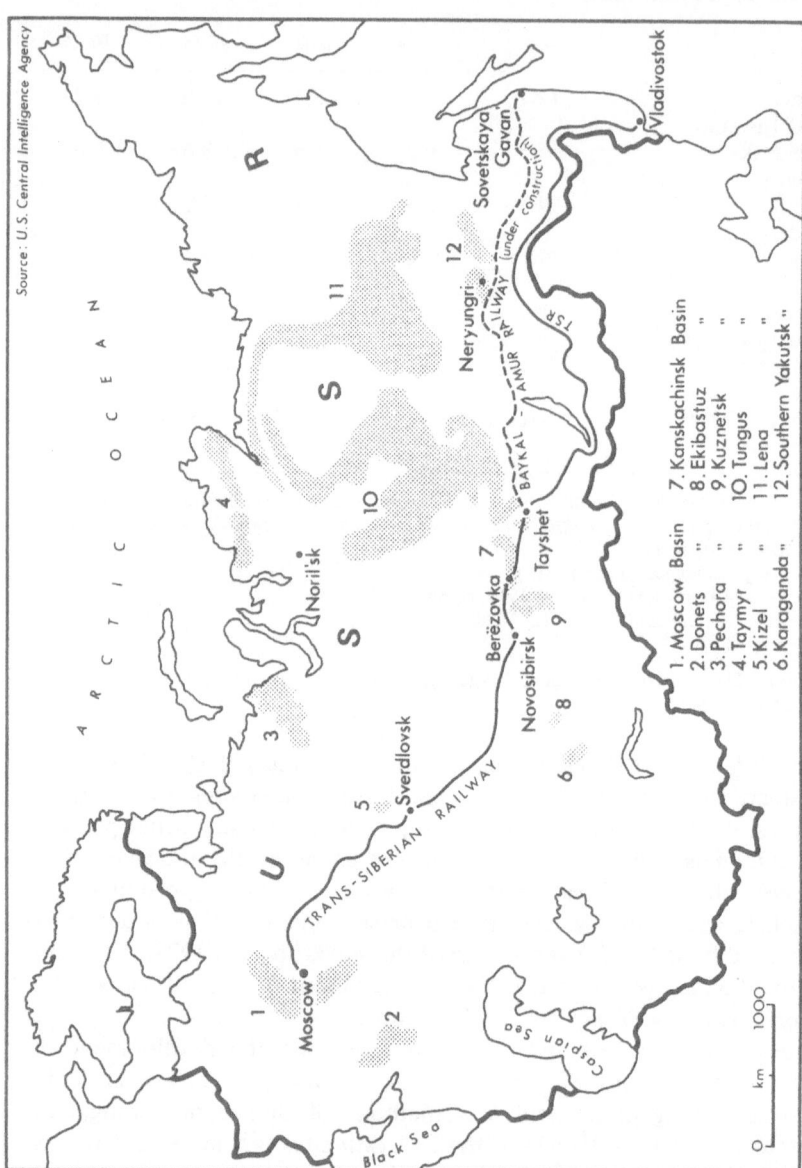

FIGURE 5.1   Coalfields of the USSR

who downgraded production targets during the 1970s and now expect that the Basin will at best maintain its present output levels. In economic terms, future investments in the Basin offer a poor rate of return when compared with Eastern coalfields, or oil and gas development, and this is likely to be reflected in Soviet planning during the 1980s (see Table 5.6). However, strategic factors and the desire of Ukrainian planners to minimise dependence on hydrocarbon imports should inhibit the more hawkish views of Moscow.

The Moscow Basin produced an estimated 29mt in 1978, a marked decline from the level of the 1950s. Its coal is mainly lignite, with a low calorific value, high moisture content, high ash content and a tendency to spontaneous combustion. Although strip mining has been introduced during the 1970s, most mining is underground, with the highest production costs in the USSR. To add to this tale of woe, the large quantities required by electricity power stations, which take most of the output, cause severe environmental problems. Nevertheless, its position ensures that it will be a major source of energy for the Moscow area during the 1980s, although present levels of production may be difficult to maintain.

Pechora Basin. This coalfield, which lies deep within the Arctic Circle, was developed during the Second World War to provide coal for unoccupied North-European Russia. Although permafrost and geological conditions make mining difficult, and haulage distances are great, it has nevertheless remained an important source of supply to the region, particularly of coking coal for the steel industry. In 1978, production amounted to 28mt of high-ash, high calorific-value coal and this level should be maintained, or increased slightly, during the 1980s.

Urals. As with Pechora, these fields came into prominence during the Second World War, providing the energy base for the Urals metallurgical and industrial complex. Although it continues to provide a valuable local source of energy, production has declined from the 1960s, with Siberian and Kuznetsk coal being shipped in to replace it. In 1978, production of a variety of coals amounted to 45mt, and output will continue to decline in the 1980s. However, this may be offset during the late 1980s by exploitation of extensive lignite deposits in the Southern Urals.

The Karaganda Basin produced 48mt of high-ash hard coal in 1978, mainly from underground mines. About 60 per cent of production is steam coal, which is used within Central Asia and Kazakhstan, although some is transported to the Urals and European Russia. Most coking coal is also shipped to the Urals, although the amount declined during the 1970s as an indigenous steel industry was built up. Output from Karaganda stagnated during the late 1970s and is unlikely to raise by more than 10–20 per cent during the 1980s.

Kuznetsk Basin. With its 1978 production of 145mt – of which 40 per cent was coking coal – this West Siberian field is the second largest producer within the USSR, and will become the first by the end of the 1980s. Its rapid expansion since the Second World War rests upon its high-quality, thick-seamed deposits, which cost only half as much to mine as those of the Donetsk. Although local industries have recently been developed, most Kuznetsk coal is transported by rail to the Urals and European Russia – where, despite journeys of up to 2000 kilometres, it is competitive with local coal. Some proceeds even further, to Western Europe, and accounts – together with Kuznetsk coal sent east to Japan – for 15 per cent of USSR coal exports.

Although output stagnated in the late 1970s, ambitious plans are in hand to greatly expand production, making the Basin the lynchpin of the future Soviet coal industry. In 1979, construction of thirty-five new surface mines and thirteen new underground mines was announced, with the intention of trebling production to 485–500mt per annum by 1990, compared to previous official projections of 200–350mt per annum for that year. Over half of output would then come from strip mines, compared to 29 per cent in 1978.

Even before this announcement, there were considerable doubts about the ambitious targets set for the field. Construction delays and other factors have greatly increased the capital cost, whilst further problems have been encountered with recruitment and productivity of labour, and railway and electrical transmission capacity for westward transport of the coal or its energy. As these factors will continue to be important, the official targets are unlikely to be met, although substantial increases in output will undoubtedly occur.

Ekibastuz Basin. With Kuznetsk, this field accounted for much of the 1970s increased production, and the achievement should be repeated in the 1980s. Although its sub-bituminous coal has a high ash content and medium calorific value and requires special boilers for its combustion, its extraordinarily thick seams – up to 40m – and the fact that it can be surface mined make it the cheapest coal in the USSR, and a more attractive investment than many oil and gas deposits. In 1978, production amounted to 57mt, most of which was sent by rail to power stations in Kazakhstan, the Urals and European Russia. Output is planned to increase to 115mt per annum in 1985 and 170mt per annum in 1990, a large part of which will come from the giant Bogatyr mine, with a capacity of 50mt per annum. To ease the pressure on railway capacity, most of the extra production will be burnt in 4 000MWe power stations, each consuming 16–17mt per annum of coal. These will be built over the next ten years, with the first scheduled for the mid-1980s. Most of the power will then be transmitted over a new 1 500kV line to the Moscow region. However, problems have already been encountered with the innovative cable technology required, and the line is unlikely to be ready much before 1990, if it is constructed at all. The effects of this on production may be exacerbated by a rising ash and waste content in the coal (up to 46 per cent by some reports) and shortages and poor quality of mining equipment and labour. For these reasons, the CIA sees 150mt per annum as the most realistic production target for 1990.

Kansk-Achinsk. These huge lignitic deposits are the great hope of the Soviet coal industry. Seams are up to 50m in thickness and can easily be surface mined at a cost similar to that of Ekibastuz. However, production failed to expand as much as was hoped during the 1970s, and amounted to only 32mt in 1978. The reason appeared to be the low calorific value, poor transportability and unusual chemical characteristics (making special boilers necessary) of the coal; lack of local demand due to cheap hydro power; and distance of alternative markets. Soviet plans envisage an output from Kansk-Achinsk of up to 300mt per annum by 1990, with the coal burnt in eight to ten 6 400MWe power stations for transmission by 1 200kV, or even 2 400kV, power line to Western Siberia and European Russia. However, the enormous technical problems involved appear to be giving Soviet planners second thoughts about such a scheme. One alternative is to beneficiate the coal – a 25mt per annum plant is under construction – and to transport it west, perhaps to the Kuznetsk, displacing local coal which could then be transported to European Russia. Another is to build slurry pipelines, either to transport raw lignite or upgraded coal. Finally, coal-conversion plants could be built, and a 1.2mt per annum liquefaction facility is under construction at Krasnoyarsk. However, an estimate by the Russian SFR planning commission states that the conversion of 350mt per annum of Kansk-Achinsk coal would require an investment of 20 billion roubles – equal to the entire assets of the oil and gas industry in 1976. Given the technical obstacles, and financial requirements of all the means of using coal from this region, it seems unlikely that production will more than double until the end of the 1980s, or that the deposits will become a major element in the Soviet coal industry until the end of century.

South Yakut Basin. These deposits of high-quality coal have been

made accessible by the construction of the Baikal–Amur railway. A
Japanese-financed project is to exploit 400mt of reserves around the
newly-built coal town of Neryungri, with an output of 13mt per annum.
Of this, 80 per cent will be coking coal, of which 5mt per annum will be
sent to Japan in repayment of loans. However, Soviet journals have
reported that Sibgiproshakt, the Siberian mining construction agency,
has greatly under-estimated the effects of harsh terrain and severe
climate, and that infrastructure has been delayed by lack of co-
ordination between the bodies concerned. According to one estimate,
the cost has quadrupled from 1.2 to 4.5 billion roubles, and the intended
start-up date of 1983 will be postponed by several years.[2]

Other coalfields of the USSR include those around Lvov in Western
Europe – which are mainly burnt in power stations feeding the Mir
(Peace) transmission network to Eastern Europe – and deposits which
are found along the Trans-Siberian Railway, and mined for local use,
notably around Irkutsk. Enormous coal resources are also known to
exist in the Lena, Tungusska and Taymyr Basins of Northern Siberia,
but there is at present only very limited production for local needs. The
enormous capital investment required, the difficulties of transport, and
the opportunities in Kansk-Achinsk and other parts of Siberia make it
unlikely that these areas will be tapped on a large scale until the twenty-
first century.

Organisation

Coal production in the USSR is under the general direction of the
Ministry of the Coal Industry, which employs about 2 million people, of
whom 600 000 are underground workers (see Table 5.4). The complex,
and inefficient, structure of the industry was simplified in the 1970s, with
the creation of autonomous regional corporations – reporting directly
to the Minister or indirectly via Republican Coal Ministries, especially
in the Ukraine – as the link between coal enterprises and the centre.
Research and planning is generally concentrated in Moscow, as with the
Central Scientific and Economic Institute of Coal (TsNIEIUgol) which is
responsible for strategic planning. Integration with other energy sectors
(which include its main markets) is achieved through Gosplan, the state
planning agency. Ministries which greatly affect the fortunes of coal
include those for Electric Power; Power Machine-Building; Geology
and Railroads. Despite recent changes, failure of co-ordination still

seems to be a common problem, as with the development of the South Yakut coalfield. Another example reported in the Soviet press occurred in 1978, when shortfalls from the Donets Basin threatened the achievement of national production targets. Although the Kansk-Achinsk field was already producing more that its customers could immediately use (due to slow conversion of boiler capacity) they were nevertheless told to increase their coal orders by 8.5 per cent. As a result, between August and October 1978, ninety loaded trains were stranded because of lack of handling capacity at customer plants, with the end result that the lignite deteriorated to the point of unusability.[3]

Although similar stories have occurred within the mining industry itself, in general it has successfully installed new capacity and increased productivity at a steady rate (although not always up to plan targets) during the last twenty years. As in the West, this has mainly been due to mechanisation and the increased proportion of national output coming from surface mines.[4]

However, by the late 1970s progress was slowing – the yearly growth of surface mined output fell from 6 per cent per annum in the early 1970s to 3.5 per cent per annum in the early 1980s – and it became clear that increased investment in huge, capital-intensive strip mines and fully-automated underground faces was necessary. These changes have been, and will continue to be, hampered by technological deficiencies in the mining equipment industry, with the result that machinery is unavailable or unreliable and has a poor performance. Thus in 1981 over half of Soviet rope-operated shovels had a bucket capacity of less than 5 cubic metres, and there were chronic shortages of drilling rigs, bucket-wheel excavators and walking draglines.[5] Similar problems occur in the electronics industry, where a high proportion of output is diverted to military purposes. Although these difficulties may be partially eased by imports from free market countries, they may inhibit the growth of coal production for some time.

Labour productivity may be a further constraint on output (see Table 5.4). The Soviet mining force has a high average age – in 1976 only a quarter of workers were less than 35 years old – and recruitment of younger miners has proved difficult, causing labour shortages in some coalfields. One reason for this is the difficult working conditions in some Soviet pits and/or the harsh environmental and social conditions in the frontier coalfields of Siberia and the Arctic. Another factor is a shortage of consumer goods on which wages – which are high compared to other industrial occupations – can be spent, which also restricts the use which can be made of bonus and incentive payments.

TABLE 5.4   USSR: labour force and productivity, by major coal basin

	1970	1971	1972	1973	1974	1975	1976	1977	1978[a]
	Thousand workers[b]								
Total employment	876.6	848.4	815.6	790.8	773.0	767.0	784.0	792.1	800
Donetsk (underground)	470.2	454.3	434.5	422.2	416.7	422.4	438.6	446.9	451
Moscow (underground)	43.5	41.3	39.0	36.4	33.5	31.4	29.5	28.6	29
Kuznetsk (underground)	119.0	117.0	114.0	112.0	110.3	107.8	108.9	110.0	111
Kuznetsk (open pit)	13.1	13.0	12.9	12.7	12.6	12.5	12.8	12.8	13
Karaganda (underground)	45.6	45.1	43.8	42.7	41.4	40.1	40.1	40.6	41
Pechora (underground)	31.5	29.9	29.1	28.4	27.6	26.9	27.6	28.1	28
Other	143.8	147.8	142.3	136.4	130.9	125.9	126.5	125.1	126
	Metric tons per year								
Total output per worker	702.0	747.6	795.6	836.4	877.2	904.8	901.2	903.6	905
Donetsk (underground)	459.6	478.8	500.4	519.6	526.8	524.4	510.0	496.8	470
Moscow (underground)	NA	NA	NA	NA	962.4	1 000.8	964.8	945.6	NA
Kuznetsk (underground)	836.4	889.2	943.2	992.4	1 044.0	1 113.6	1 141.2	1 155.6	NA
Kuznetsk (open pit)	2 148.0	NA	NA	NA	2 775.6	3 042.0	3 120.0	3 262.8	NA
Karaganda (underground)	810.0	NA	NA	NA	1 040.4	1 095.6	1 124.4	1 126.8	NA
Pechora (underground)	681.6	732.0	772.8	810.0	847.2	900.0	933.6	949.2	NA
Other	1 253.7	1 372.8	1 480.4	1 580.4	1 730.3	1 852.3	1 911.6	1 977.6	NA

[a] Estimated.
[b] As reported in the No. 4 issues of *Ugol'*, 1971–8, and estimates and projections of the labour force and civilian employment in the USSR, *Foregin Economic Report*, No. 10, 1976, p. 31, Department of Commerce.

SOURCE   US Central Intelligence Agency, 'USSR: Coal Industry Problems and Prospects', (1980).

In theory, Soviet miners are the recipients of extremely good fringe benefits, to compensate for the dangers of their work. During the 1970s their working hours were reduced, and are usually lower than those of other industrial groups, while retirement takes place at the age of fifty. However, it appears that high production targets and an absence of independent trade unions cause a number of health and safety problems, which have created discontent in some pits. The example of strikes by Polish miners to solve similar grievances may well lead to an increase in such actions within the USSR.[6]

Markets

As in the West, an increasing amount of coal in the USSR has been burnt to generate electricity, and the power-station market now takes over 50 per cent of all coal mined. Under the 10th Plan (1976–80), twenty-one new solid-fuel plants were anticipated, compared to fifteen under the previous Plan. Some consideration has also been given to the conversion of existing oil-fired stations into coal-fired. By the end of the 1980s, coal should provide about 50 per cent of all electricity generated, compared to 37 per cent today. However, certain factors may prevent the more optimistic plans for coal-fired electricity. Over one third of power-station fuel consumption goes to heat production in combined heat and power stations – Thermal Electric Centres (TET) in the USSR – and, because of their high energy efficiency, these will form an increasing percentage of newly commissioned capacity. However, because of their predominantly urban situation – to supply large concentrations of housing and industry – coal is at a disadvantage to other fuels, particularly natural gas, because of the greater amount of pollution which it causes, in part because of the poor quality of anti-pollution equipment. Further problems have occurred in the construction engineering industry, particularly with regard to the provision of special boilers needed to burn certain types of coal and the conversion of power stations from oil/gas to coal. However, in the medium–long term, the effects of these factors may be mitigated by the introduction of new coal-combustion technologies, such as MHD, where the USSR is presently the world leader in its research and development. The minemouth stations of Soviet Asia will also continue to be purely electricity generators – known as condensing electric stations (KES) – although their development may be hampered by increased concern about environmental effects.

The other main market for Soviet coal is coke-making, both for the iron and steel industry and as a feedstock for chemical manufacture (although during the last two decades much of this market was lost to oil and gas). In 1978, 182mt (25 per cent of total output) was used for these purposes, of which 82mt was from the Donets, 60mt from the Kuznetsk, 19mt from Karaganda and 17mt from Pechora. Although the Asian USSR accounts for most coking-coal production, coke-making plants are concentrated in the European USSR, making the westward transport of approximately 50mt per annum necessary. In coming years it is likely that these westward flows, particularly from the Kuznetsk, will increase, to offset decline in Donets production, but this may be compensated for by the end of the decade as coke plants are re-located in the East.

Another market for coal within the USSR is for domestic and industrial heating. Soviet interest in coal conversion was limited during the 1970s, but it is now seen as a partial solution to the problems of transporting Siberian coal. Several plants to liquefy and gasify lignite have been built in the Moscow and Kansk-Achinsk Basins.

The USSR's coal exports in 1981 amounted to 22mt – about 8 per cent of world coal trade. Most of these went to COMECON countries, particularly East Germany, Bulgaria and Czechoslovakia, and also to Finland. Exports to Japan and the European Communities were also significant in the 1970s, but were temporarily abandoned in 1981 as a result of tight domestic supplies. Some increases are planned for the future, particularly from the South Yakut field. However, delays in the development of this scheme, together with the likelihood of a tight internal energy market for much of the decade, makes it unlikely that the USSR will increase its share of world coal trade, and may well cause it to decline. The USSR also imported 3.8mt of coal, mainly from Poland, for its European regions in 1981.

Infrastructure

The mis-match between coal reserves and coal markets makes the cost of transportation a major constraint on coal output (see Table 5.5). In fact, shortage of railway cars was one of the official reasons given by the Soviets to explain the decrease in production during 1978. With major problems to be solved in the technology of high-voltage transmission lines and slurry pipelines, a large percentage of coal will still be transported by rail during the 1980s. East of the Urals, the railway system appears to be generally adequate, and the planned double-

TABLE 5.5 USSR: delivered cost of coal (rubles per metric ton of standard coal equivalent[a])

Origin	Destination						
	Leningrad	Moscow	Gor'kiy	Ulyanovsk	Sverdlovsk	Novosibirsk	Krasnoyarsk
Donetsk	25.2	22.0					
Kuzentsk	21.0	20.4	18.1	17.1	13.3	9.2	
Ekibastuz		18.6		14.1	10.1		
Kansk-Achinsk				18.0	13.0	8.0	5.0

[a] *Planovoye khozyaystvo*, No. 11, 1977, p. 146. It is assumed that cost data are for 1976.

SOURCE US Central Intelligence Agency, 'USSR: Coal Industry Problems and Prospects', (1980).

tracking of the Trans-Siberian Railway, together with construction of
the Baikal-Amur link, should eliminate bottlenecks and open up new
routes. However, production increases may be held back by slow
railway construction and shortages of equipment, as has been the case in
the past. West of the Urals, the system already appears to be
overburdened in several areas, and may be severely stretched during the
1980s.

Environment

Although information as to the environmental effects fo the Soviet coal
industry has been generally sparse, there is every reason to believe that
they have been widespread and damaging – perhaps to an even greater
extent than in the West.[7] Subsidence is frequent in the Donets Basin and
other underground mining areas, whilst air, land and water quality
seems to have been damaged in most coalfields. It seems likely that
similar problems of reclamation of surface mines to the Western USA
occur in parts of the Asian USSR, such as Karaganda and Ekibastuz.

Most Soviet coals also have a high sulphur content – in the European
USSR over 1.5 per cent – but the 10th plan envisages removal of only 81
per cent of particulate matter and gaseous emissions. In these
circumstances, the problems of dust, hydrocarbons, sulphur dioxide and
acid rain must be serious in many areas, especially those which make use
of low calorific value coal (needing greater quantities for a given heat
output). Severe problems in this respect can be anticipated in the Kansk-
Achinsk field, which can experience almost permanent temperature
inversions – up to depths of three miles – during the winter. Disposal of
the 70 mt per annum of ash and sludge which are created by Soviet
power stations must also pose difficulties, with some solutions – such as
dumping into rivers at Ekibastuz – merely transferring the problem to
other areas. Finally, the enormous water requirements of coal-fired
plants, and mines themselves, can degrade scarce water supplies,
particularly in semi-arid areas such as the Asian steppes.

Coal and Energy Policy

Interpretations of the likely trends in the Soviet energy system in the
1980s are confusing, especially with regard to the availability of oil and
gas.[8] However, although it appears that adequate reserves exist – the

CIA having modified its more pessimistic analyses of the late 1970s – it is also clear that Soviet planners face major problems in ensuring their future exploitation. Particular concerns include the worsening imbalance between Eastern energy supplies and Western markets; the difficulties of discovering new and maintaining production from old, sources of oil and gas and the wisdom of building too many nuclear power plants near to population centres in the European USSR. More generally, Western studies have found 'no evidence of any systemic shift in the USSR toward a less energy-demanding economic structure' which might reduce the future demand for coal and other fuels.[9]

Given these fears, it is probable that the recent emphasis on coal production will be maintained during the 1980s and that output will expand. However, the dismal record of the coal industry under the 1976–80 Plan – output only rose from 712mt in 1976 to 716mt in 1980, rather than a targeted 805mt – illustrate the scale of the problem. This is reflected in the less ambitious targets of the 1981–5 Plan, which envisages an increase in coal output of 5 per cent per annum, to 785mt by 1985. Delays in opening up new capacity, or difficulties in increasing markets and transporting coal to them, may lead to resources being transferred to other energy industries, to open up presently inaccessible oil and gas deposits (see Table 5.6 for comparative costs). The fact that coal expansion requires huge investments in Siberia may also become

TABLE 5.6 USSR: capital investment per ton of standard fuel by source of energy[a] (rubles)

Coal	
Kansk-Achinsk	9.6
Ekibastuz	8.2
Kuznetsk (open pit)	27.8
Karaganda	30.3
Donetsk	64.3
Moscow	89.7
Natural gas	
Tyumen'	29.4
Central Asia	35.6
Oil	
Tyumen'	29.2

[a] Source: *Planovoye khozyaystvo*, No. 6, 1975, p. 66.
The Soviet data include an interest charge on the stock
of reproducible fixed assets.

SOURCE US Central Intelligence Agency, 'USSR: Coal Industry Problems and Prospects', (1980).

entangled in the general political debate on the best distribution of resources between the Western and Eastern USSR.

One implication which can be drawn from these conclusions is that the USSR is unlikely to become a major coal exporter, especially to OECD markets. At most, exports might double to 50mtce per annum by 2000 – as forecast by the WOCOL Report – whilst the implication of more pessimistic analyses – such as that of the CIA – is that they will not significantly rise above present levels, while imports to the European USSR, particularly from Poland, may increase.[10] Confirmation of these trends was given in the early 1980s when Soviet exports to most Western countries were halted, due to domestic and other problems.

CHINA

With the USA and USSR, China is the third in the triumvirate of great powers which dominate the world coal industry.[11] Despite the political and economic vicissitudes of the 1970s, output has increased in every year since 1968, to stand at 644mt in 1982, although high amounts of ash and limited washing facilities reduced its energy content.

With recoverable reserves of 990btce and additional resources of 13394btce, the Chinese industry can increase production for many years without fearing the exhaustion of its coal deposits.

Geography

Over 80 per cent of Chinese coals are bituminous, with generally high ash and sulphur contents. Deposits are widely distributed over the country, although the largest concentrations occur in the North and North-Eastern regions (see Table 5.7 and Fig. 5.2). In 1979, the leading coal production provinces (fields of more than 10mt per annum output in brackets) were Shanxi (Datong, Yangquan), Liaoning (Fushun, Fuxin), Shandong, Henan (Pingdingshan), Heilongjiang (Hegang, Jixi), Anhui (Huaibei, Huainan), Hebei (Kailuan, Fengfeng), Jilin, Sichuan and Yunnan. An estimated 65 per cent of national reserves are contained in Shanxi and Nei Monggol provinces, with four others – Anhui, Shandong, Henan and Gansu – having a further 16 per cent between them. New investment is being concentrated in the Northern and Central regions, particularly the provinces of Shanxi and Shandong, where, the cost of developing new mines is 50–90 per cent

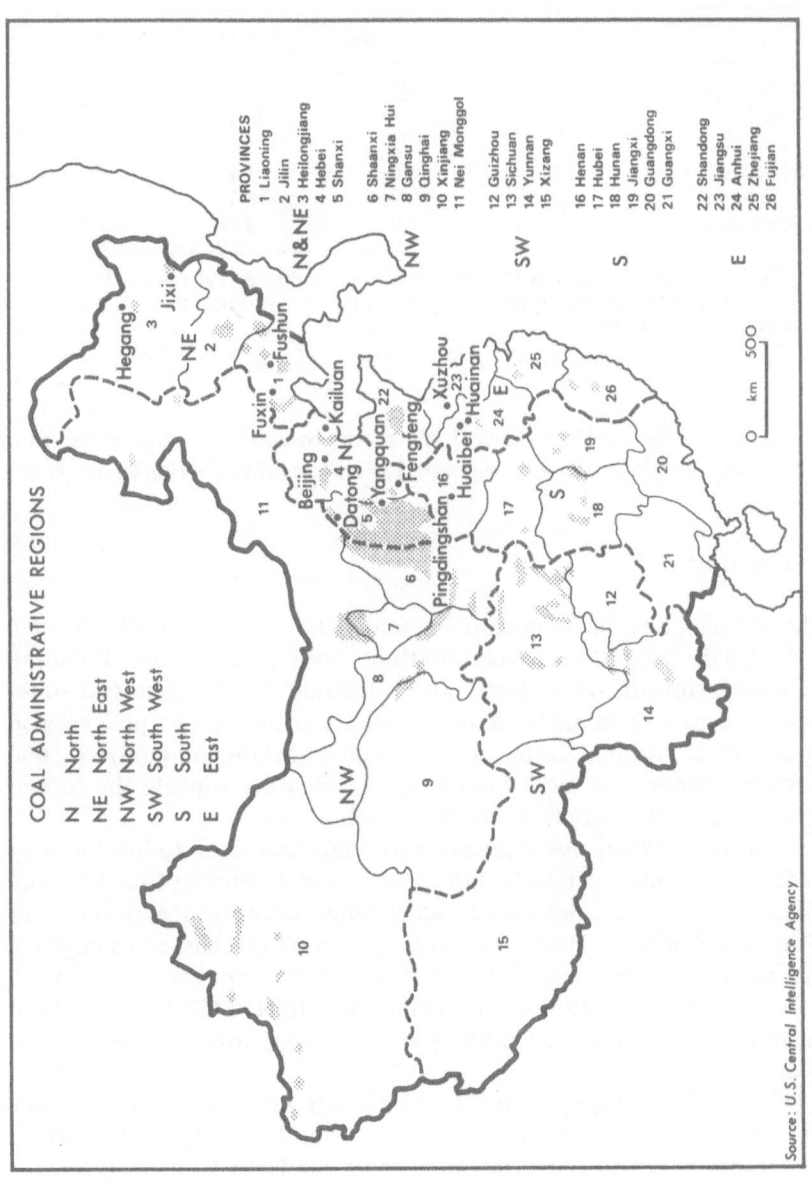

FIGURE 5.2 Coalfields of China

TABLE 5.7   China: regional distribution of coal production and reserves, 1975[a] (per cent)

	Production	Reserves
Northeast	22	2.7
North	28	70.1
Central	22	4.0
South	10	1.2
Northwest	9	18.7
Southwest	9	3.3

[a] Data on regional output in 1978 have not been estimated, and the disturbances within the coal industry in 1976 and 1977 would distort the distribution of output.

SOURCE   US Central Intelligency Agency, 'Chinese Coal Industry', (1979).

less than in the south.[12] Several major coal bases of 20–50mt per annum capacity are planned with associated chemical and electricity complexes.

Organisation

Most coal production in China is under the auspices of the Ministry of Coal which – in collaboration with the State Planning Council – formulates strategic and operating plans. Some 600 state-owned mines – administered by Provincial Coal Bureaux – produced 56 per cent of national output in 1981. Local government controls 2 000 smaller mines and 1 700 local mines – which are outside the formal planning system – are run directly by communes.

During the early 1980s greater autonomy was given to local mining administrations and new bodies – such as the Southwest China Energy Resources United Corporation, which aims to develop coal resources in Yunnan, Guizhou, Hunan and Guangxi – were established to improve co-ordination between Coal, Rail and other ministries. A China National Coal Development Corporation (CNCDC) has also been formed to import coal technology and encourage foreign investment in Chinese coal resources.

Given the background of economic and political uncertainty, as well as the devastating effect of the 1976 Tangshan earthquake – which temporarily wiped out production from the large Kailuan field – the success of the Chinese coal industry in raising production during the 1970s was a creditable achievement. However, it was clearly strained when new capacity of 30mt per annum was being built, and increased production targets were met in part by a higher percentage of waste and lower grade coal. Thus, between 1977 and 1978 an increase in tonnage production of 20 per cent produced only a 4–10 per cent increase in the calorific value of mined output. Expansion was also due to the opening up of small mines and the more intensive exploitation of larger ones. According to the CIA: 'This growth formula will no longer work; large-scale investment in new, large modern mines can no longer be put off. China lacks the technological knowledge and equipment needed and, as a practical matter, cannot obtain them from foreign sources in time to achieve the plan.'[13] Although this source is subject to bias, the decline in production during 1980 and the official downgrading of the 'four modernisations' programme and its ambitious production targets, suggest that it is accurate. One result has been cancellation and postponement of many projects, and reductions in the level of machinery imports. In future, purchases seem likely to be tied to coal exports as a means of paying the debts incurred. Export contracts may also be linked with direct investments or loans, as is already the case with Japan and Spain.

At present, foreign technology is being concentrated on the development of mechanised longwall retreat faces, of which over 100 were in operation by 1979. Some 8mt per annum was also produced by hydraulic mining in five pits. The proportion of coal won by these mechanised means is expected to rise from 33 per cent in 1979 to over 50 per cent by 1985, although this target may be somewhat ambitious. At Tatung, output from mechanised faces working in near-perfect geological conditions has been only 20 per cent of that achieved with similar equipment in the USA, largely because of management and maintenance problems, and bottlenecks in coal haulage. Elsewhere manual mining methods, involving firing and hand-loading the coal on to conveyors, and supporting the roof by timber props – shortages of which have been a major constraint on production in the 1970s – will continue into the 1990s. As a result, poor safety conditions – causing an estimated 2 500 – 5 000 fatalities and 100 000 injuries a year, as well as many cases of pneumoconiosis – are expected to continue.[14]

To take maximum advantage of available equipment, many large mines have been reconstructed or re-organised, whilst some machinery has been transferred from one coalfield to another, as with British equipment moved from the Datong field to others in Shandong. Efforts have also been made to manufacture local copies of some foreign equipment which has been imported on a 'one-off basis'. Although the mining-machinery industry has had a creditable performance in the past, it may well be hampered by the conflicting demands of other extractive industries and shortages of skilled labour. This, and the high economic and social value of land in most parts of China, may also hamper plans to expand surface mining – which presently accounts for only 5 per cent of output – in the coming decade. Most foreign visitors to China have reported that the mining workforce – which numbers over two million – is skilled and disciplined, but that collieries are generally overmanned, with little concept of productivity measurement. However, this is compensated for by the high level of capacity utilisation, which often approaches 24 hours a day in some pits.

Markets

Although coal's share of Chinese primary energy supply has declined from 90 per cent in the early 1960s to 70 per cent in 1980, it should continue to be the main fuel until the end of the century. At present, coal consumption in China is similar to that of the rest of the world in the 1950s, with a diversity of markets.

In 1979, 39 per cent of coal output was for chemical and general industrial use, 21 per cent for residential use, 18 per cent for electricity generation, 17 per cent for coking and 5 per cent for railway use. Environmental and transport problems are expected to concentrate a greatly increased share of consumption on electricity generation, particularly at new minemouth power stations, and perhaps coal-conversion plants. However, in the short–medium term, economic problems will favour a development of hydro sources to avoid the import of foreign power-station technology. Plans to increase coking-coal production to meet steel industry targets have also been reduced. Although some expansion should take place, a lack of domestic sources of high-grade coking coal means that imports will be necessary, at least until adequate washing and blending facilities have been developed within China.

China exported 7mt of coal in 1981, mainly to the nearby countries of Japan and North Korea. Contracts have been signed to supply Japan with at least 10mt per annum by the mid-1980s, a large proportion of which would be coking coal. A second large contract with Spain envisages exports of up to 12mt per annum by 1987, whilst exports are also planned to Hong Kong and other South Asian and West European countries. However, despite official optimism, investment cutbacks, transport problems and difficulties with oil production will probably limit exports to 20mt per annum until the 1990s, although they may expand substantially beyond that point.

Infrastructure and Environment

Transport problems will be a major constraint on domestic, as well as export, coal consumption during the 1980s. They include insufficient railway track and rolling stock capacity, long hauls from mines to consumers, in many cases over 2000km – and a shortage of port facilities. The problems are particularly acute in South China, where coal shortages have occurred because of lack of capacity on North–South links (on which coal comprises 60 per cent of freight traffic). Some of these problems will be eased, but not completely solved, by large investments to cut out bottlenecks – although these have been affected by re-appraisal of the 'four modernisations' programme. New railway lines are under construction, or planned, from central coalfields to southern industrial centres. If new minemouth power stations are built, less coal will also need to be transported. Finally, new facilities are being provided at Chinese coal ports, particularly at Qinhuangdo and along the Yangtse river, which should benefit both internal and export traffic.[15] At Qinhuangdo, a $160 million project should treble coal capacity to 30–40mt per annum by 1985, and the port should handle the bulk of Chinese coal exports.

As in the USSR, it is only recently that evidence on the environmental effects of coal mining and use in China has become available, but the government has now admitted that it is a severe problem. In fact, the level and distribution of coal exploitation, together with an absence of environmental controls, suggest that the effects have been extremely deleterious. Efforts have been made to close down polluting industries – much of which will be attributable to coal-burning – in Peking and other tourist areas, while reports of subsidence and other environmental damage at mine sites are common.

Coal and Energy Policy

The original projections of the 1976–85 Plan for coal production of 1bt per annum by 1990 and 2bt per annum by 2000 have been abandoned for more realistic targets – which are 8–900mt per annum by 1990 and 1.3bt per annum by 2000. Although output will undoubtedly grow steadily during the 1980s, the problems which the industry faces – illustrated by the sacking of the Coal Minister in early 1980 and official admissions that 'Coal and energy have been the weak links in the national economy' – and the re-appraisal of the modernisation programme throw even these forecasts into doubt.[16] The course of events will partially depend upon developments in the oil industry, where production appears to have reached a temporary plateau, forcing China to renege on some export contracts with Japan during 1980. Coal exports may be particularly affected, either because domestic consumption rises as a result of energy shortages, or because it is used to replace oil in domestic uses, with the latter being exported to take advantage of its high foreign currency value. For all these reasons, Chinese plans to become a major coal exporter in the 1980s, and the WOCOL estimate of 30mtce per annum exports by the year 2000, could be over optimistic.

EASTERN EUROPE

All the East European members of COMECON, and Yugoslavia, share four characteristics which are fundamental in shaping their energy policies: (a) all are, or will shortly become, net energy importers; (b) all have coal deposits which are being exploited with determination; (c) at present, most of their imports are obtained from the USSR but, although these supplies will be maintained, in future they will have to turn to non-COMECON sources for incremental supplies; (d) coal imports are likely to be easier to obtain than other fossil fuels.[17] In the past, these countries have traded large amounts of coal amongst themselves, both directly and indirectly, through the Mir (Peace) electricity grid, and this situation is likely to continue.

POLAND

Poland is by far the largest producer of coal in Eastern Europe, with an output of 163mt of hard coal and 35.6mt of lignite in 1981.[18] Its 41.3mt of exports in 1979 was second only to the USA, and accounted for 19 per cent of the world coal trade, although sales fell to 15mt in 1981 as a result of political and social unrest. Recoverable reserves of 30.6btce and additional resources of 91.2btce will allow existing production to be maintained or increased in the future.

Geography

The Upper and Lower Silesian Basins account for almost all Polish hard-coal production, with the Upper Basin predominant. The coal is of good quality, with low ash and sulphur content, generally occurring in thick seams which are little disturbed by faulting, although rockburst and gas hazards are frequent. It is mainly used as steam coal: some 15–20 per cent of output goes for metallurgical purposes. Since the Second World War, large production increases have occurred, concentrated in the Upper Basin where pits have been rationalised and investment concentrated on highly-mechanised mines which are amongst the most modern in the world. However, working out of more accessible deposits means that increases in production will not be possible in this zone, so that future investments will be concentrated at greater depths and in more difficult geological conditions at its northern edge. Targets for the Upper Basin were production of 232mt per annum by 1985, and 300mt per annum by 2000, and for the Lower Basin maintenance of the present 4.5mt per annum output. These are upper limits and have been adversely affected by economic and industrial problems.

Lublin Basin. This deposit is conveniently situated to serve markets in Eastern Poland, and possibly for export to the USSR. Production from seven new mines is planned to reach 7mt per annum by 1985, 16mt per annum by 1990 and 46mt per annum by 2000. Most of this output will be consumed at minehead power stations. However, geological and infrastructural problems, and capital shortages, may cause considerable delays in the field's development.

Lignite. Existing lignite fields near to the East German frontier supply minehead power stations. New deposits have been discovered in Central Poland, especially in the Kanin, Betchatow and Szczerow areas, with estimated reserves of 12 billion tonnes. Production is planned to reach 40mtce per year by 1990, which will also be burned in large minehead power plants.

Organisation

Overall responsibility for coal production in Poland lies with the Ministry of Mining, situated at Katowice, in the Silesian coalfield, rather than the capital of Warsaw. Coal exports are handled by an independent organisation, Weglokoks. Almost all Polish hard-coal production is from underground mines, and heavy investment during the 1970s has meant that a high percentage of output is derived from highly-mechanised longwall retreat faces.[19] As a result, productivity in 1979, at 4.1 tonnes per manshift, was the highest in Europe, although it later fell as a result of new working conditions. Further increases are expected when the highly efficient Lublin mines enter full production, with productivity levels an estimated five times those of Silesia. This success has enabled Polish mining technology and expertise to be sold extensively overseas, including Yugoslavia, Nigeria and Venezuela. Despite this, lack of capacity and manpower problems has meant that output has not kept pace with demand in the late 1970s.

One solution increasingly favoured by Polish authorities is the use of foreign financing for new developments, with repayment in the form of long-term coal contracts. A major precedent was set in 1980 when an Austrian consortium provided $300 million for expansion in Upper Silesia, to be repaid in coal deliveries of up to 1.5mt per annum for twenty years from 1984 onwards. A West German credit of DM1.2 billion, part of which is financing the Lublin coalfield, was also linked to future Polish coal exports. Further deals would ensure that the eight new mines planned to supplement the existing sixty-six are open by the mid-1980s, but the already high level of Polish hard-currency debts may make their arrangement difficult.

Polish mineworkers have been relatively well paid – up to three times the industrial average – and were conspicuously absent from the industrial disturbances of the 1970s. However, grievances over health, safety and other working conditions boiled over in the summer of 1980, when mine strikes forced the authorities to grant industrial concessions and to allow the existence of free trade unions.[20] Productivity fell and, although it rose after the 1981 military clampdown, tacit opposition by mineworkers has kept it below the high levels achieved in the 1970s. Whatever the outcome, problems will also result from inadequate recruitment, as in 1979, when less than a third of 50 000 new workers remained in the industry.

Markets

Markets for Polish coal are more diversified than in most other countries. In 1978, 70 per cent of electricity generated was derived from coal, and this market consumed 52mt of hard coal and 40 mtce of lignite. Coal-fired capacity is scheduled to rise from 21 000MWe in 1979 to

51000MWe by 1985, with further construction by the year 2000. Most of this new capacity will be in the form of minemouth power stations.

Some 25mt of hard coal was used for metallurgical and coke purposes in 1978, and about 3mt of coke was exported. Construction of a large iron and steel complex in Southern Poland should increase this market considerably during the 1980s, although this may be offset by government policy aimed at reducing the amount of energy – intensive, low-value production in the national economy.

Direct industrial and domestic use of coal accounted for approximately 40 per cent of coal consumption in 1978, of which over half is used in industrial boilers. This market is forecast to drop substantially during the 1980s to 10 per cent of coal output, being replaced by coal-conversion products such as electricity and gas, which may account for approximately 40 per cent of coal consumption by the end of the century. West German interests are financing the first large-scale gasification plant at Libiaz, Silesia, which should produce 1 billion cubic metres per annum of gas by the mid-1980s.

In 1979 Poland was the world's leading steam coal exporter, sending 25mt to Western and 15mt to Eastern Europe. Industrial unrest reduced exports to 15mt in 1981, of which almost half went to COMECON countries, although they rose to 20mt in 1982 and higher in 1983. Even before the 1980–81 events the WOCOL projection of 50mtce per annum exports seemed optimistic and the limits on production arising from worker discontent, foreign exchange shortages and investment problems make it unlikely that exports will resume the 1979 level for many years, if ever. Political factors may also lead to a higher percentage of Polish coal exports going to COMECON nations than in the past – a trend already becoming apparent in the early 1980s.

Infrastructure and Environment

During the 1970s, coal distribution was subject to several bottlenecks, particularly rail capacity from Silesia to the central Polish cities and Baltic ports, and coal-loading equipment at the docks of Gdansk and Swinoujscie. Substantial investments are planned to circumvent these, including a circular railway line around the Silesian Basin – allowing coal to be concentrated at depots along it for outward shipment by unit train – and new lines from Silesia to the Baltic, and to connect the

Lublin field to the existing network. Improvements to ports and waterways, including construction of a barge port at Tychy, on the Upper Vistula, are also planned. Finally, the use of slurry pipelines to transport Polish exports to Central Europe has been proposed, and these may be in operation by the late 1980s.

The scale of mining and coal use in Poland also creates substantial environmental problems.[22] Although regulatory standards are comparable to those in the West, they are not always observed. Air quality in Silesia is reported to be the worst in Europe, and its improvement was one of the demands put forward by workers in the unrest of the early 1980s. Some remedial measures can be expected, and other more urgent, issues make it unlikely that environmental factors will constrain future production.

Coal and Energy Policy

Although Polish coal production increased at a steady rate of 3.5 per cent per annum in the 1970s, it seems unlikely that this will continue during the 1980s. Technical, recruitment and transport problems seem likely to occur, and new investment will be handicapped by the country's general economic problems and high degree of indebtness to the West. The opposition of miners to the military government which brutally crushed the Solidarność union in 1981 may also result in lowered productivity for some years.

Before these events a Polish study saw 240mt per annum as the maximum output which could be achieved before the 1990s and this level now seems optimistic in the short-medium term. As a result of rising domestic energy consumption, this would mean a stagnation in coal exports and make Poland a net energy importer for much of the next decade.[23] Should problems occur over the price or availability of oil imports, there may be pressure to reduce coal exports in favour of domestic conversion to liquid or gaseous fuels. This policy will also be influenced by the outcome of the debate on the Polish economy which the unrest of the early 1980s and the poor growth rates have provoked. One option is to guide the economy away from low-value, energy-intensive production and the export of raw materials to higher-value goods, of which coal-conversion products would be a part.

In conclusion, it is evident that coal production will remain the mainstay of the Polish economy and – directly or indirectly – a major source of foreign currency earnings. However, exports to free-market economies are unlikely to return to their 1979 level and may remain depressed for many years.

BULGARIA

Bulgaria produced 38mt of coal, almost all lignite, in 1981, which met a quarter of primary energy needs.[24] Recoverable reserves amount to 1.9btce, with additional resources of 1.6btce. The main deposits are the East and West Maritsa lignite fields, and hard coals in the Sofia Basin. Most lignite output is used for electricity generation, where special techniques have been developed to handle its high ash and moisture content. Imports, mainly of metallurgical coal, amounted to 6mt in 1981, mainly from the USSR, but these will diminish when the recently-discovered coking-coal deposits at Dobrovdzha are exploited.

CZECHOSLOVAKIA

Coal production in Czechoslovakia amounted to 27.3mt of hard coal and 95.3mt of lignite in 1981. Recoverable reserves are 3.4btce with additional resources of 6.5btce.

Bituminous deposits are concentrated in the Ostrava-Karvina Basin of North Moravia, an extension of Poland's Silesian coalfield. These have a low ash and sulphur content, and are of coking quality. The main lignite deposits are found at Usti-Kadan in North Bohemia, which produce 70 per cent of national output from seams of up to 20m thickness. Other deposits are exploited in West Bohemia, South-East Moravia and at Handlova, Slovakia. Over 80 per cent of lignite is from mechanised surface mines.

Electricity generation consumes half of coal output, with the rest going to industrial, metallurgical and domestic markets. Although demand will rise, it is thought that adverse geological conditions and the increasing ash content of lignite deposits limit the opportunities to increase production. As a result imports, which totalled 4.4mt in 1981, mainly from Poland and the USSR, will rise in the 1980s, although still being met from within COMECON.

EAST GERMANY

East Germany is the world's leading producer of lignite and the 1981 output of 267mt was its main energy source. Recoverable reserves stand at 7.5btce, which will not allow any significant production increases. The main lignite deposits are in the Cottbus and Halle-Liepzig areas, with others at Magdeburg extending into West Germany. Underground mining has been discontinued and output now comes from thirty-seven surface mines.

Almost all output is used in minemouth power stations, which produce 80 per cent of East German electricity, although a small proportion is briquetted and coked. The environment problems of burning such large quantities are considerable and may encourage a move to gasification. Lignite output has stagnated in the 1970s, and increasing output to 300mt per annum will require the opening of twenty-one new mines over the next decade. This will be especially difficult because of the increasing depth of overburden over new reserves. As a result hard coal imports, which amounted to 6.5mt in 1981, from Czechoslovakia, Poland and the USSR, will significantly increase in the 1980s and East Germany may be forced to enter the world markets.

HUNGARY

After declining in the late 1960s, and stagnating in the 1970s, Hungary's coal industry is set to expand from the 25.3mt of lignite and 3mt of hard coal which it produced in 1981.[25] With recoverable reserves of 1.5btce and additional resources of 1.76btce – both mainly lignite – its targets should prove to be attainable.

Lignite deposits are found in three main areas: the Transdanubian Basin in north-west Hungary, the Varpolita Basin a few miles to the south, and the Borsod-Nogrod fields of north-east Hungary, all of which experience mining problems due to excess water and methane. Despite this, a major expansion programme is taking place in the Transdanubian basin, where almost $2 billion is being invested in four new and two reconstructed mines to supply local power stations when in full production by the end of the 1980s. Hungary's sole hard coalfield is in the Mecsek district, near Pecs in South Hungary, where seven pits tap deep and tectonically-disturbed deposits. A third of the output is suitable for coking.

Responsibility for the industry lies with the Ministry of Heavy Industry, and operations are carried out by the Hungarian Coal Mines Trust, which operates some fifty mines. Sales are equally divided between the electricity and industrial boiler markets, but the former is likely to grow at the latter's expense in the future. Forecasts suggest that output will reach 30 mt per annum by 1990 and 50mt per annum by 2000, and the absence of other energy sources makes it likely that this target will be achieved. Existing levels of coal imports – 1.7mt in 1981 – from other COMECON nations, mainly Czechoslovakia and the USSR, are likely to continue and may even increase in the short–medium term. In the long term, however, energy shortages within COMECON may reduce the availability of these supplies, forcing Hungary on to the world market in the absence of other indigenous energy sources.

ROMANIA

Romania's coal industry has always been overshadowed by its much larger oil industry, but the decline of oil reserves has brought it into prominence. In 1981 output was 37.9mt, mainly of lignite, and ambitious plans have been formulated to expand production, taking advantage of recoverable reserves of 413mtce and additional resources of 520mtce.

Lignite mining is concentrated in the Arges and Mortu fields, and bituminous coals in the Jiu Valley. Most of this production was consumed in nearby power stations. Official plans call for an expansion of production to 75–5mt per annum by 1985, mainly in the form of surface-mined lignite for electricity generation. However output has been consistently below targets and this situation is likely to persist. Particular problems include industrial unrest – a major strike shut down the Jiu field in 1977 – lack of equipment and poor co-ordination. To some extent, shortfalls may be made up by an increase of coal imports from the 2.2mt of 1982. Some of this will come from the USSR, but a contract for some 2mt per annum has also been signed with the US company, Island Creek Coal.

YUGOSLAVIA

Although it is not a full member of COMECON and is classified as a developing country by the World Bank, Yugoslavia's political and economic system, and the structure of its coal industry and markets, make it more akin to other East European nations than the West. Production amounted to 51.6mt, mainly of lignite, in 1981, and provided a third of primary energy needs. With recoverable reserves of 8.7btce and additional resources of 2btce, there is ample scope for increasing output in the future.

The largest deposits are found in the Kosovo Basin, which contains lignite seams of up to 100m thickness. Output was 9.9mt in 1979 and is planned to reach 100mt per annum by 1990. The Kolobara Basin produced about 13mt in 1979, and projected to reach an output of 20mt per annum by 1990. Expansion is also planned for the Tuzla Basin – from 13.5mt in 1979 to 25mt per annum in 1990 – and the Klac Basin – from 1.2mt in 1979 to 10mt per annum by 1985. The new Sibovac field should also be producing 5mt per annum by 1985. In total, output is projected to reach 80mt per annum in 1985 and 210mt per annum by 1990.

Over 70 per cent of coal production is burnt in minemouth or nearby power stations. As the construction of new oil-or gas-fired stations is banned, this market should rise from 30mt in 1979 to 70mt per annum in 1985 and about 150mt per annum in 1990. Some 9 000MWe of new capacity is planned for the Kosovo Basin alone. Hard-coal imports for metallurgical purposes amounted to 3.7mt in 1981, and will increase during the 1980s.

Although Yugoslavia has an easily-mined reserve base for these ambitious plans, the coal industry's disappointing performance in the 1970s suggests that their implementation will be difficult. Problems of co-ordination and lack of capital have not yet been solved, and the country's economic performance may deteriorate in the 1980s. The environmental problems of mining and burning such large quantities will also be considerable.

NORTH KOREA

North Korea produced 48mt of coal in 1981, much of it anthracite, and mainly from mines in South Pyongyang or North Hamyong provinces. Small quantities are exported to Japan and China, although the scale of these are constrained by periodic shortfalls in domestic coal supplies, particularly for electricity generation.

With recoverable reserves of 534mtce and additional resources of 4.4btce, North Korea plans to expand coal output to 70–80mt per annum by the mid 1980s, although similar attempts at rapid expansion were unsuccessful in the 1970s. Most mines are already very deep and still using backward technology, so that these official targets will require considerable investment to be achieved.

VIETNAM

The Vietnamese coal industry was greatly affected by war, but in 1979 produced 6.2mt, mainly from the Hanggai deposits. Exports amounted to 0.7mt, mainly to Japan and France. With recoverable reserves of 150mtce and additional resources of 700mtce, expansion of both production and exports are planned. An important element in these goals will be investment in new mines in Quang Ninh province, improved coal-loading facilities at ports and a coal gasification plant for fertiliser production.

6 The World Coal Trade

Coal is a widely-distributed resource which is expensive to transport and is usually of low value. Accordingly, most production is consumed within national boundaries, and in 1981 only 276mt – 10 per cent of world hard coal output – was internationally traded. During that year the main exporters were the USA (102mt), Australia (51mt), South Africa (29.9mt), the USSR (22mt) and Poland (15mt), and the main importers Japan (78.2mt), France (27.4mt), Italy (18.4mt), Canada (14.8mt) and West Germany (11.3mt). Several countries – notably Canada, West Germany and the USSR – were both substantial importers and exporters. In regional terms, coal imports were dominated by Western Europe (113.5mt), Japan–South Korea (88.1mt) and Eastern Europe (29.4mt).[1]

COKING COAL

Just over half of world coal trade is in the form of coking coal, a basic raw material of the steel industry for which no substitutes have, until recently, been available. In the past, the trade was primarily in premium low-volatile coals, largely derived from the USA and, to a lesser extent, Poland and the USSR. Modern blending techniques have allowed much greater amounts of medium/high-volatile coals to be used in coke production which, together with the desire of importers for diverse sources of supply, has allowed new reserves to be exploited. In Europe, it has allowed the use of domestic coking coals to be maintained, with their relatively poor quality being offset by continued low-volatile imports. In Japan, it has led to the replacement of premium coking coal imports by medium/high-volatile supplies from Australia and Canada, where mines are geared to the export market, and linked with consumers by long-term contracts and, increasingly, equity investments and loans. This contrasts with the USA and other traditional sources, where most coking-coal output is sold to domestic users and exports usually take place under short-term agreements.

In the early 1970s, record steel production and the impact of rising oil prices caused considerable increases in coking-coal shipments and prices. In the late 1970s and early 1980s, world recession led to a stabilisation or fall in shipments and prices, with a decline of investment in new mines and facilities. However, most observers believe that steel production, and coking-coal demand, will resume its upward trend in the 1980s and beyond. The International Energy Agency (IEA) has forecast that coking-coal imports will reach 193mtce per annum by 2000, while the WOCOL Report predicts that import demand could reach 260–300mtce per annum by that year.[2] Although some growth can be expected under most circumstances, continuing global economic problems and the development of new steel-making techniques which do not require coking coal – such as direct reduction – may hold back demand. Improved blending techniques may also blur the distinction between steam and coking coals, so that the increased volume of metallurgical coal shipments is not matched by increases in value. It would also lead to changes in the geography of coking-coal trade as more potential sources are created. Geographical changes will also be caused by the development of Third World steel production, in part replacing Western output.

STEAM COAL

Steam coal is primarily used for steam raising in electricity generation or industrial plant and has a lower value than coking coal. It is in direct competition with other fuels, which are usually more convenient to use, and demand is therefore extremely sensitive to price differentials and transport costs. The volume of trade fell during the 1960s and early 1970s, but began to grow after the oil price rises of 1973. Most steam-coal movements have been from Poland to Eastern and Western Europe, with the latter also being served, to a lesser extent, by the USA and South Africa. In the late 1970s, Japan and East Asian countries imported growing quantities from Australia and South Africa, and in the early 1980s reduced Polish exports and other factors allowed these countries to expand their share of the European market.

Several studies have predicted that world steam-coal trade will grow dramatically, with the IEA predicting imports of 337mtce per annum by 2000 and the WOCOL Report 300–680mtce per annum for the same year. As now, demand would be concentrated in Western Europe and East Asia, although South America and, possibly, the Southern USA will also be of importance. The main exporters will be Australia, the USA and South Africa, with China possibly replacing Poland as the main Communist supplier. The relatively low value of steam coal will lead to a markedly regional trading pattern. Greater use of long-term contracts and direct investment by consumers in mines is also likely.

As with coking coal, the achievement of the more optimistic forecasts of steam-coal demand growth will depend upon world economic conditions in coming decades. In addition, they assume that existing price differentials between steam coal and competing fuels will be roughly maintained and that the introduction of larger bulk carriers will reduce marine freight charges, neither of which may be justified. Finally, the lower value, and less secure markets, of steam coal may make it difficult to co-ordinate the simultaneous investments in production, transport and user facilities which are necessary to provide a 'coal chain'.

ORGANISATION

The organisation of a rapidly-growing world coal trade, particularly in steam coal, will be a major enterprise. Even in 1980–1, the disruption of Polish exports as a result of political and social unrest caused severe problems for several importers and affected market conditions everywhere.[3] A great extension of brokerage services to match buyers and sellers will be necessary, while the provision of 'coal chains' will lead to the setting up of public or private bodies to co-ordinate developments in several countries. Major problems will also be encountered in raising the enormous sums of capital required for coal expansion, particularly for individual projects. According to the WOCOL Report, $1 000 billion (in 1978 prices) will be required to meet OECD demand by 2000, a large proportion of which will be spent on projects connected with exports or imports.

The signs are that these functions will be taken over by subsidiaries of the multinational oil companies, which perform similar services in the world petroleum market. They are already major producers and reserve-owners of coal, and several act as selling agents for other mines. They – or other subsidiaries – are also major shipping operators and, in future, may be significant consumers of coal for synfuels production, where they are presently spending large sums on research and development. Particularly prominent companies are BP and Shell, both of which have interests in Australia, Canada, South Africa, the USA and Western

Europe, and Exxon, which is opening up large coking-coal reserves in Colombia and has extensive interests in the USA.[4] To a lesser extent, mining organisations – such as Ruhrkohle (West Germany), Charbonnages de France and the National Coal Board (UK) – are also developing similar international interests.

Although the financial and manpower resources of these bodies will aid the development of world coal trade, the trend is likely to prove as controversial as the oil companies' present domination of the world petroleum markets. In particular, concern will mount – it is already significant in the USA – that the companies may utilise their position to control coal–oil differentials and to establish high coal prices, accepting if necessary a reduction in volume growth in return for improved profit margins. Although most official studies in the USA have found that this has not been the case, the issue may well be of great political significance by the 1990s and after.

AUSTRALIA

Although only partial exploration has yet taken place, Australia's coal potential is great.[5] Recoverable reserves amount to 36.3btce, with additional resources of 612btce. These deposits will allow the 1982 output of 101.8mt hard coal and 38mt of lignite to be enormously expanded by the end of the century, by which time Australia may be the world's leading coal exporter.[6]

Geography

The main mining areas of Australia (See Fig. 6.1) are:

New South Wales where mines in the Sydney Basin produced 64.9mt of hard coal – 60% of national output – and accounted for 45 per cent of reserves. Deposits are mainly low-volatile bituminous, with low sulphur and medium ash content, and often suitable for coking. Most coal is produced in underground mines, but the extent of surface mining is expected to increase in the 1980s. Production has been concentrated near to the industrial markets and ports of the Sydney and Lower Hunter River–Newcastle areas, but future increases will come from inland deposits, particularly in the Upper Hunter River, where exceptionally thick seams lie near to the surface.[7]

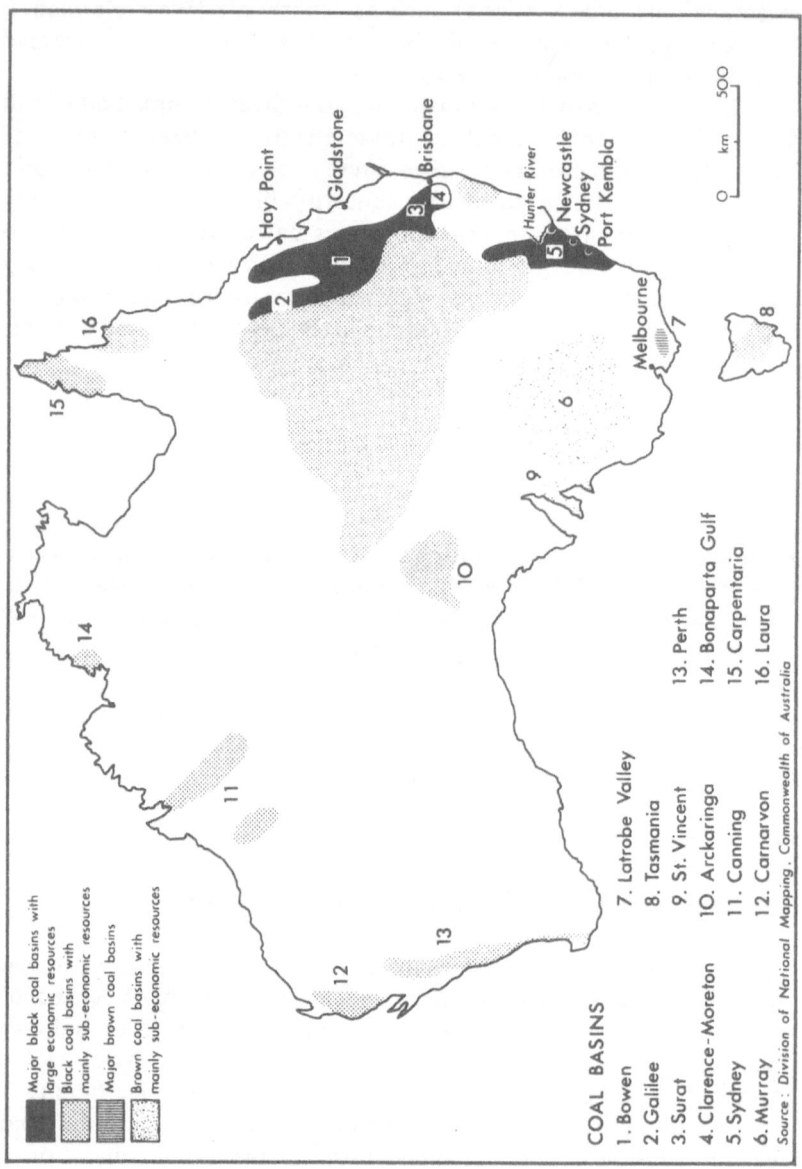

COAL BASINS

1. Bowen	7. Latrobe Valley
2. Galilee	8. Tasmania
3. Surat	9. St. Vincent
4. Clarence-Moreton	10. Arckaringa
5. Sydney	11. Canning
6. Murray	12. Carnarvon
13. Perth	
14. Bonaparta Gulf	
15. Carpentaria	
16. Laura	

Source : Division of National Mapping , Commonwealth of Australia

FIGURE 6.1   Coalfields of Australia

Queensland mines had a total output of 36.8mt of hard coal in 1982, almost all from surface mining. Most of this came from the Bowen Basin in South-East Queensland, particularly those areas adjacent to the coast. These deposits are shallow and mainly comprise medium–high rank bituminous coals, with low ash and sulphur content, although in their eastern extremities tectonic activity has converted them to anthracite. The remaining production comes from high-ash steam-coal deposits adjacent to coastal markets, particularly the Surat and Clarence–Moreton basins, South-West of Brisbane. Although coal deposits are known to occur in much of interior Queensland, sufficient reserves exist in Eastern Queensland to ensure that this area will dominate production until the next century.

Victoria contains most of Australia's lignite reserves, and the 33mt which was produced in 1981 accounted for almost all of national output. Production is concentrated in the Latrobe Valley area of the Gippsland Basin, in South-East Victoria, where seams of 90–230m thickness, with 10–140m of overburden, occur. Other deposits are exploited in the Anglesea area of the Otway Basin, South-West of Melbourne. All production is worked and used by the State Electricity Commission of Victoria. Small amounts of hard coal are also mined in the Collie Basin of Western Australia (where output should reach 5mt per annum by the end of the 1980s), at Leish Creek in South Australia, and Fingal in Tasmania.

Organisation

The Federal Department of Minerals and Energy has overall responsibility for coal development, including the power to veto coal export agreements. Much power also lies with the State Departments of Mines or Minerals and Energy, who provide tax incentives for mine and infrastructure development. Two boards, which bring together government and producer interests – the Joint Coal Board of New South Wales and the Queensland Coal Board – also exercise some administrative duties. Scientific and technical work has been supported by a coal-production levy since 1977 and is mainly undertaken by the Australian Coal Industry Research Association. The Australian coal industry is diverse, with many different production companies. Some 20 per cent of hard coal, and all lignite output, is captive production for public utilities, notably the New South Wales Electricity Commission and the State Electricity Commission of Victoria, and steel companies, particularly the Australian Iron & Steel Co. The largest private Australian producers are Broken Hill Proprietry, Coal & Allied Industries and Thiess. Numerous foreign companies also have Australian coal interests, especially oil companies (for example, Shell, BP, Exxon, Arco, Total), mining companies (for example, RTZ, NCB, Ruhrkohle, Utah) and Japanese organisations.[8] Usually a 50 per cent local ownership is required, and several projects have been stalled in recent years until the requirement has been met. Major developments in which foreign interests are playing a major role include the German Creek, Oaky Creek and Blair Atholl mines of Queensland, and the Oakbridge development of New South Wales.

The modern nature of Australian mines, the export orientation of many, and the favourable mining conditions of most, make the industry efficient and capital-intensive.[9] Over half of hard-coal output is from surface mining, and this percentage should continue to increase until at least the 1990s. Most underground mines which remain will continue to use the present room-and-pillar methods, although both longwall and shortwall (a technique developed in New South Wales) faces can be expected to increase.

Although industrial relations in the coal industry have been historically difficult, especially during the late 1940s and early 1950s, the siuation has eased in recent decades. In New South Wales, days lost as a result of disputes amounted to only 3.3 per cent in 1976–7, compared to 18.3 per cent in 1949–50. Output stood at 12.4 tonnes per man-shift in New South Wales and 16.4 tonnes in Queensland in 1979, one of the highest levels in the world. However, extensive strikes in the early 1980s re-opened doubts in some foreign markets, particularly Japan, as to the reliability of Australian exports. Some observers are also worried that shortages of skilled labour may also hamper the coal industry's expansion, although some remedial action has been taken by both the Federal and State governments.

Markets

In 1982, utility coal-burn of 25mt hard coal and 35mt of lignite provided 75 per cent of Australia's electric power. As economic planning for the 1980s is based upon cheap and abundant supplies of coal-derived electricity, demand from this market can be expected to increase rapidly until the end of the century. Some eighteen new coal-fired generating plants are planned to double installed capacity to 1987.

Much of this demand, especially in Queensland, will be for 'in-house' electricity production by mineral processors. In particular, five coal-powered aluminium smelters are planned to exploit local bauxite deposits, and Australia is expected to be the world's major exporter of this commodity by the end of the 1980s. The WOCOL Report forecasts that the electricity market will require 43mtce by 1985 and 107mtce by the year 2000.

The metallurgical market accounted for 7mt in 1982, mainly from captive mines owned by steel producers. Although demand is likely to grow in the future, this market may decline in percentage terms as coal purchases for alternate uses grow more rapidly. Forecasts suggest that demand will reach 17mtce by 2000, although this depends on the extent to which the Australian steel industry is able to withstand competition from other Asian producers. Similar quantities of coal may also be required for industrial steam-raising markets, a marked increase from the 3mt which this took in 1978–9 – although the outcome will depend upon the price differentials between coal and domestic oil/gas.

In the medium–long term, many Australian coal deposits seem especially well suited to gasification and liquefaction processes. Australian states and mining companies have linked with West German, Japanese and US interests, and it is probable that at least one full-sized plant will be operating by the end of the 1980s. The most likely sites appear to be Millmerran, Queensland and the Latrobe Valley, Victoria. The Australian 'High Coal Use' projection for the WOCOL Report sees a demand for 25mtce per annum coal for synfuels production by the year 2000, whilst the 'Low Coal Use' scenario discounts this market entirely.

Coal exports amounted to 48mt in 1982, providing approximately 10 per cent of all Australian export revenues.[10] Two-thirds of these went to Japan, with Western Europe and South Asia being the other main markets. Only 20 per cent of this was in the form of steam coal, but increased trade in this commodity should occur in the future. Despite attempts to diversify markets, the willingness of Japanese buyers to outbid competitors and to sign long-term contracts will mean that they continue to take a dominant share of exports in the 1980s. One particularly significant deal which illustrates this trend is that signed in 1980 between the Japanese Electrical Power Development Corporation and the Blair Atholl mine of Queensland. This provides for the sale of 5mt per annum of coal from 1985 to 2000, and for an eventual Japanese stake of 19 per cent in the venture. The WOCOL Report forecasts that Australian exports could reach 160–200mtce by the year 2000. Although this is technically feasible, achieving this volume (especially its upper level) depends upon considerable investment and active government co-ordination and encouragement.

Infrastructure and Environment

Inadequate transport facilities, particularly railways and ports, hampered production in the 1970s. A substantial part of new mine investment funds, as well as large sums of public money, will be required for such infrastructural needs – amounting to $17.3 billion over the next twenty years. In the early 1980s, inland transport problems will be reduced by increased emphasis on the Upper Hunter Valley region of New South Wales, but after this point large investments will once again be required to open up inaccessible Queensland deposits. The existing coal ports at Newcastle, Sydney and Port Kembla, New South Wales, and Hay Point and Gladstone in Queensland should all be able to handle 120 000DWT carriers by the mid-1980s, but it still seems possible that port facilities may constrain production in the late 1980s and beyond. Lack of facilities in potential importing nations may also limit coal exports for significant periods.

As elsewhere, concern for the environment and regulatory legislation has increased in Australia during the 1970s, and looms increasingly large as an influence on coal production and use. In 1981, mining in part of the Hunter Valley field was restricted on these grounds. Further problems which will be faced in the 1980s include the movement of new mines into more arid and remote areas; the dust and noise resulting from coal transportation, particularly around the ports; and the pollution from coal combustion in power stations, which may be concentrated in certain coastal areas.

Coal and Energy Policy

There seems little doubt that Australian coal production will expand rapidly to the year 2000, both for export and a growing domestic market. However, a question remains as to the precise rate of this expansion and, in particular, whether the optimistic forecasts of the WOCOL Report will be achieved.

Even with great encouragement by Federal and State governments – which at present is forthcoming – the problem of co-ordinating the coal system, minimising environmental and health problems, and providing adequate infrastructure are formidable. Capital requirements will be immense and the industry will face competition from other industries to obtain it. In practice, much will be obtained from overseas, creating a risk of economic 'over-heating' and increased inflation.[11]

The exchange rate may also be influenced, with possible deleterious effects on local manufacturing industry. In these circumstances, it is conceivable that government policy in the future might favour a less rapid rate of expansion, with consequent effects on export levels. It is also possible that public opinion may become unhappy with Australia's role as a supplier of low-priced raw materials to the world and favour indigenous processing – in the case of coal into liquid/chemical products and energy-intensive goods – again at the expense of direct exports. Although much can change in the course of two decades, it seems likely that the euphoria of the late 1970s and early 1980s may be followed by a more sober analysis of the industry's future, particularly its rate of expansion and pricing policies.[12]

SOUTH AFRICA

South Africa produced 130.3mt of coal in 1981, reportedly at the lowest production cost in the world, due to cheap labour and good mining conditions. Although there has been some controversy about precise figures, recoverable reserves are estimated at 25.3 btce, with additional resources of 33.8 btce.[13] Production should expand steadily for both domestic and export markets until the end of the century.[14]

Geography

The Karoo Basin is the largest coal basin of South Africa (see Fig. 6.2) and is adjacent to the main markets. Deposits are mainly medium-quality bituminous, with a generally low sulphur content (0.5–1.5 per cent), located in thick seams at shallow depths. Volcanic activity has created narrow anthracite seams in Natal and a hard dolorite sill above the coal measures in much of the Basin, which can make roof control difficult.

Mining is carried out in discontinuous zones, of which the Witbank area is the most important. Some thirty collieries produced over half of national output in 1981. The other main areas are at Ermelo (four mines

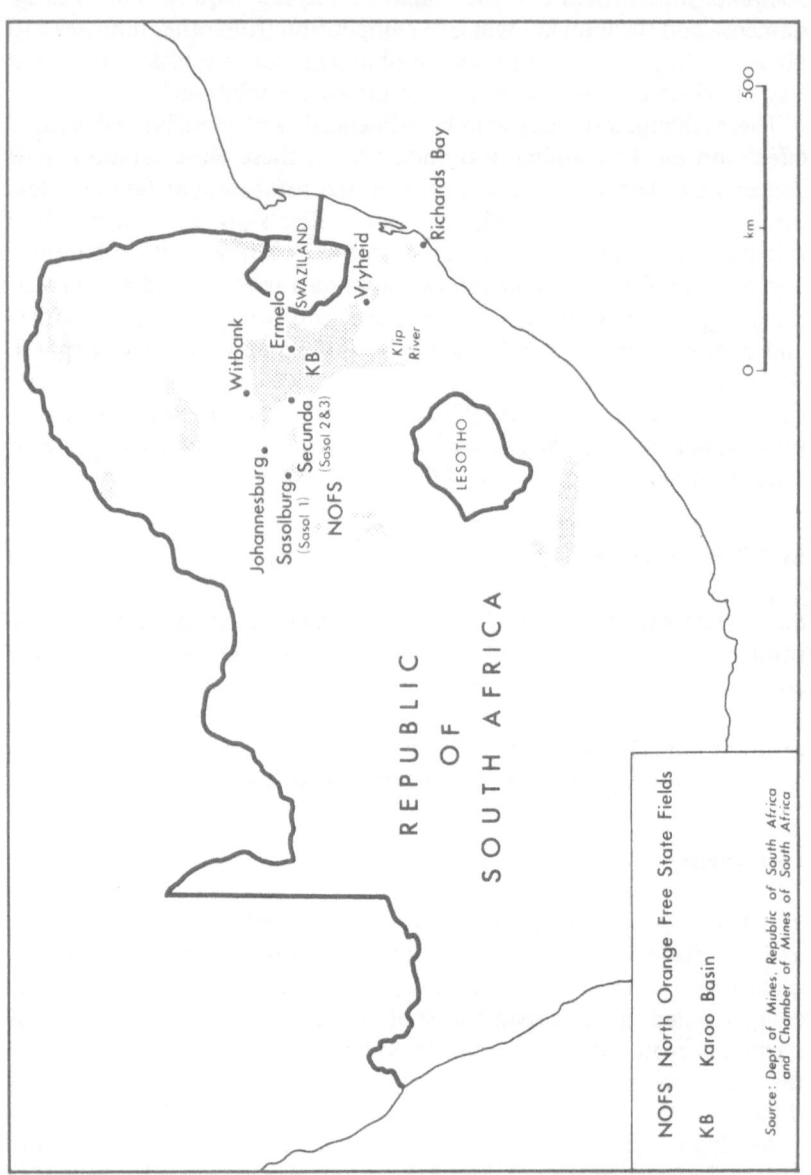

FIGURE 6.2 Coalfields of South Africa

producing 8 per cent of national output) and the Vryheid and Klip River regions of Natal (twenty-three anthracite mines producing 13 per cent of national output).

North Orange Free State. Although geologically related to the Karoo deposits, these low/medium-grade bituminous deposits around Vereeniging and Welkom are considered as a separate field. Five collieries (one in Southern Transvaal) produced 20 per cent of national output in 1981, all for coal-conversion and power-plant use.

Other fields are known to exist in Central and Northern Transvaal, Eastern Natal and Eastern Cape Province, but have as yet been little exploited. Although development can be expected in the future, particularly of those with good-quality coking coal or near to export ports, the central coalfields will remain predominant for several decades.

Organisation

Overall responsibility for coal development and production in South Africa lies with the Department of Mines. Since the 1950s, the government has operated a 'cheap energy' policy, which has involved considerable intervention in the coal market. Price and export controls have been enforced, with the result that domestic prices have lagged behind those of the rest of the world. As a result, capital has been difficult to attract, investment in new equipment has been discouraged and only the most profitable seams have been exploited. However a liberalisation of export quotas was announced in 1981 and the problem should be less serious in future.

Coal-producing interests are represented by the South African Chamber of Mines, which has some management functions, including wage and condition negotiations. As with other South African minerals, coal output is controlled by five large mining finance houses. Two of these – Anglo American and General Mining – controlled 67 per cent of total coal sales in 1978.[15] The other large producer is SASOL, whose mines provide coal for its coal-conversion plants. However, several foreign companies – notably BP and Shell – have been investing in coal exploration and development, and should become a major force by the end of the decade.

In 1978, some 84 per cent of total production was from underground mines, but the opening of several large surface mines should reduce this percentage to about 66 per cent by the mid-1980s. Almost all coal is extracted by room-and-pillar working, using conventional mining methods. The number of longwall faces will increase, but their use is limited by the shallowness or discontinuity of many seams. Output per man-shift was 7.2 tonnes in 1977, when the workforce was 120 000.[16] Economic pressures and a legal requirement to avoid subsidence has meant that recovery rates are low – in some cases, only 10 per cent. Concern has been expressed that this sterilisation of reserves might constrain future production. Despite extremely low wages (often below the official poverty level) and sometimes poor safety standards and living conditions, industrial relations in most mines have been good. However, the politics of apartheid in South Africa – reflected in the largely black labour force and largely white management – may affect future production, either by violence as in the 1980 bomb attacks on the SASOL complex, or through shortages of skilled workers due to job reservation for whites. It is expected that the entry of overseas interests into the industry may effect changes, foreshadowed by the 1980 wage rises in BP-related collieries as a result of pressure in the UK and other countries.

Markets

Most coal produced went to electricity generation, which consumed 52mt in 1981. Most of this went as captive supplies to the minemouth power stations of the semi-public utility, ESCOM, usually at very low prices – an average of R7 per tonne in 1979. The consequent cheap tariffs have meant that electricity demand has grown at an average rate of 9 per cent per annum in the last two decades – one of the highest rates in the world – and forecasts suggest that this will continue during the 1980s. Orders have been placed, or are expected, for five 3 600MWe plants to be constructed at minemouth sites during the next few years.[17]

South Africa is unique in that a large amount of coal is converted into liquid and chemical products. In 1979, the SASOL 1 plant of the South African Coal, Oil & Gas Co. at Sasolburg took 6mt of coal from its adjacent colliery in the North Orange Free State coalfield. The SASOL 2 and 3 plants at Secunda, in the Karoo Basin, will have a combined coal demand of 27mt per annum, most of which will be supplied by the adjacent Bosjesspruit colliery, making it the largest underground mine in the world.[18] Further demand will be provided by chemical plants using coal as feedstock, as with the PVC and methanol plants which two leading chemical companies, AECI and Sentrachem, are building.

Coke ovens also took 7mt of coal in 1979, of which 6.5mt went to the South African Iron & Steel Corporation (ISCOR). Demand is expected to grow steadily during the 1980s, with coke-oven consumption forecast at 11–13mt per annum by 1990 and 20–25mt per annum in the year 2000. Other minor domestic markets at present are South Africa Railways – 2.2mt in 1978 – and industrial/domestic consumption. The former market is expected to decline as coal-burning locomotives are phased out, but the latter should increase substantially.

Exports – mainly of steamcoal to EEC countries – increased to 29.9mt in 1981, largely due to the expansion of the Richards Bay coal terminal on the Indian Ocean coast. Lack of port facilities and use of official export quotas will limit exports to approximately 30mt per annum until the mid-1980s, when a further expansion of Richards Bay should allow them to reach 48mt per annum. At this point, South Africa is likely to be the world's leading exporter of steam coal. The WOCOL Report suggests that export levels will reach 55–100mtce per annum by 2000, with the latter figure as an extreme upper limit. However, fears of too rapid exploitation of high-quality reserves and political uncertainties, both domestic and international – such as the refusal of some countries, such as Sweden, to buy South African coal – may inhibit future growth.

Infrastructure and Environment

Internal rail and road facilities are adequate to meet existing levels of demand and should present no constraints on increased production. Coal exports, however, are hampered by the lack of good harbours on both the Indian and Atlantic Ocean coastlines of South Africa. Richards Bay is the only deepwater port near to the coalfields and has a maximum capacity of approximately 80mt per annum, which should be reached by the 1990s.[19] Coal could be exported through Mozambique ports, but political considerations make it unlikely that large amounts will use this route.

The environmental problems of coal production in South Africa are dwarfed by those of other types of mining and seem to have attracted little public attention. One impact which has so far been absent in South Africa is subsidence, as mineral rights do not allow the surface to be lowered. However, the increasing depth of production, together with higher recovery rates, is likely to change this situation in the future. Although some environmental controls on coal production and use can be expected during the 1980s – expecially if companies seek to exploit the high-quality deposits in protected areas, such as the Kruger Park National Reserve – it seems unlikely that they will be a major constraint on production.

Coal and Energy Policy

South Africa obtains a higher percentage of its energy needs from coal than any other non-Communist country – about 75 per cent in 1981. Although this is partly due to the availability of cheap coal, strategic considerations of minimising the use of imported oil has also been important. This has been particularly true of the SASOL plants, which have been the only commercial users of coal liquefaction technology for many years.

The construction of SASOL 2 and 3, together with an extensive coal-fired power-station programme, means that prospects for growth in domestic consumption of coal are bright, and the WOCOL Report suggests that the level may more than double to 148–173mtce per annum by the year 2000. The planned expansion of the Richards Bay terminal should also allow exports to grow steadily in the 1980s, although further major investment will be necessary to sustain this during the 1990s. The introduction of higher export quotas has also ended the debate over the strategic benefits of preserving coal reserves for future domestic use, rather than allowing them to be exported.

Finally, all assessments of the South African coal industry must take into account the domestic and international pressures which are working on the apartheid system. Major political and economic upheavals in the coming years may well have direct, and indirect, effects on the level of coal production and sales. Many importers will also be reluctant, on political grounds, to import South African coal at all, whilst others will minimise their dependence on this source.

CANADA

In the past, Canada has been a net importer of coal, but during the 1970s increased exports from its Western states reached the same level as imports to the East. In 1981, production was 40mt, providing approximately 9 per cent of primary energy needs. Recoverable reserves of 4.7btce and additional resources of 36.6btce are expected to lead to considerably increased output and exports in the future.[20]

The *Maritime* coalfields of Nova Scotia and New Brunswick were the first to be worked in Canada, but output declined to 3.1mt in 1980, mainly coking coal from the three underground mines of the Cape Breton Development Corporation (Devco). A $400 million investment programme will tap undersea deposits of the Cape Breton field, where production is planned to reach 5mt per annum by the late 1980s.

The *Plains* deposits include the lignite fields of Saskatchewan, which produced 6mt in 1980, and the sub-bituminous deposits of Alberta, whose output reached 10.5mt in 1980, mainly from the Manalta Coal Co. Both deposits are mainly used for electricity generation.

The *Western Cordillera* fields of British Columbia and Western Alberta produced, respectively, 10.2mt and 5.9mt in 1980. After rapid growth in the 1960s, expansion slowed in the 1970s but is expected to resume as these states contain the bulk of Canadian coal resources. Output is mainly from surface mines, although one underground hydraulic mine is operated, with Kaiser Resources and Fording Coal the main companies.

Responsibility for the Canadian coal industry is divided between the Federal Ministry of Energy and state governments, which usually have a Ministry of Mines. There is considerable controversy over their respective powers on energy matters, which may affect future production. The industry is generally capital-intensive, making use of large surface mines, and with US, Japanese and South Korean investment in Western Canada.

Utilities account for the bulk of coal consumption in Canada, which amounted to 38mt in 1980. Expansion of coal-fired generating capacity is planned in most states and is expected to be especially rapid in Alberta, where use of alternative fuels is banned and electricity demand will grow from the oil-shale industry. The WOCOL Report projects that this market will reach 50–65mtce per annum by 2000, compared to an official forecast of 70–85mt per annum, although these may be prevented by the environmental problems of sulphur-dioxide emissions and acid rain. It also forecasts that use of metallurgical coal will rise from 7mt in 1979 to 15mtce per annum (WOCOL) and 10–12mt per annum (official) in 2000, on the grounds that the Canadian steel industry,with access to iron-ore deposits and strong markets, is less vulnerable than those of other OECD countries. As at present, most coal for this market will be imported from the Eastern USA. The WOCOL Report also sees industrial use reaching 11–20mtce per annum and synthetic fuel demand 6–21mtce per annum by 2000.

Coal exports in 1981 stood at 16mt, almost all of metallurgical quality, and mostly to Japan, although its percentage share – 76 per cent in 1979 – declined slightly from the mid-1970s. Other markets included South Korea, Brazil, West Germany, Greece and India. The size and accessibility of Western coking-coal reserves is expected to produce a rise in exports, which the WOCOL Report projects at 27–47mtce per annum by 2000. Another study by the Canada West Foundation predicts that the total will reach 36mt per annum by1990.[21] The imbalance within Canada between Western reserves and Eastern markets will be partially rectified in future by rail–Great Lakes shipping routes, which were used for the first time in 1979. As a result, 1981 imports of 14.8mt are expected to remain at roughly their present level with the WOCOL Report projecting a figure of 9–17mtce per annum for 2000. However, a large increase in intra-Canadian coal trade may reduce the amount of coal available for export.

Transport bottlenecks may constrain future production, especially in Western Canada, although the Federal government is making large investments to overcome them. Limited rail capacity, especially through narrow Rocky Mountain valleys, is the main problem and will be expensive to overcome. Rail links between coal reserves in North-East British Columbia and Prince Rupert have been costed at $250–315 million. Ports seem less of a problem, with Vancouver expanding to a 30mt per annum capacity by the mid-1980s from the 14mt which it handled in 1979. Construction of a coal terminal at Prince Rupert is also proceeding.

Environmental factors may be a more significant constraint. Much of existing, and future, output in the Western Cordillera is from ecologically-fragile, sub-arctic mountain zones, which often have competing recreational or hydrological uses. The problem of acid rain from sulphur-dioxide emissions has also become serious in Eastern Canada and, although derived more from the USA than Western Canada, may eventually result in mandatory flue-gas desulphurisation technology for power stations.[22]

The Canadian government aims to be self-sufficient in energy by the late 1980s, and increased coal production and consumption will clearly be an important element in achieving this target.[23] Opportunities for sales of Western Canadian coking coal to Japan and Eastern Canada should be readily available, although domestic consumption of steam coal may be hindered by the 'cheap oil' policy of the Federal government. The WOCOL Report predicts that total production could reach 92–159mtce per annum by 2000, while the Canada West Foundation forecasts that output could reach 86mt per annum by 1990.

Western expansion on this scale would require substantial investment by Japanese and other foreign investors which – with associated technology transfers – the Federal government appears keen to encourage.[24] However, considerable problems of transport capacity, environmental impact, shortages of skilled labour and competition for capital from other energy projects remain and may slow the growth of output for many years.[25]

NEW ZEALAND

New Zealand produced 2.2mt of coal in 1981, mainly from the Waikato field of North Island, but also from the West Coast and Otago–Southland fields of South Island.[26] Most of this was used for electricity generation, although a small quantity was exported. Reserves of 162mtce and additional resources of 2.2btce allow scope for future expansion, which has been endorsed by the 1980 Energy Plan. This envisages production of 4.8mt per annum by 1990, of which 1.1mt per annum will be exported, mainly to Japan.

JAPAN

Japan is the world's leading coal importer, with the 77.8mt of 1982 representing a quarter of world coal trade. Domestic output has fallen from 55mt in 1961 to 17.6mt in 1982, but recoverable reserves of 1.1btce will allow this total to be stabilised until the end of the century.[27] A third of domestic production is from the Ishikari coking-coal field, on Hokkaido. Coking coal is also mined at Miike, Western Kyushu, while bituminous and sub-bituminous steam coals are exploited at Chikuho, Kyushu and Kushiro, Hokkaido. All output is from underground mines, with an average depth of over 600m and generally disturbed geological conditions, giving high production costs. Government policy is to maintain present output levels, and a Coal Mining Rationalisation Corporation has been set up to achieve this.

Metallurgical users accounted for about 87 per cent of Japanese coal consumption, and 97 per cent of coal imports are coking coal. The WOCOL Report projects that demand in this market will reach 84–86mtce per annum by 1990 and 86–92mtce per annum by 2000. However, a slight decline in coking-coal demand in 1978–80, together with signs that some bulk steel-making may be transferred to Third World countries, suggest that these figures are optimistic.

Coal was once extensively used for electricity generation, but economic and environmental factors brought a decline in the 1960s. In 1979, sixteen, mainly small, power stations were coal-fired and accounted for about 8 per cent of coal consumption. Under the 1979 energy programme, the state-backed Electric Power Development Corporation (EPDC) will construct ten new coal-fired stations, raising coal's share of electricity output from 3.5 per cent in 1979 to 10–13 per cent by 1995. Industrial use will also rise substantially, with the cement industry alone increasing its demand from about 1.6mt in 1979 to 7mt in 1985. The WOCOL Report projects that demand from these markets will reach, respectively, 57–72 and 7–12mtce per annum in 2000. Conversion of coal into liquid or gaseous fuels may also be commercial by 2000, although this is more likely to be carried out adjacent to coal deposits, particularly in Australia. A joint enterprise of several Japanese companies, the Japan Coal Liquefaction Development Co., has interests in the US SRC–2, EDS and H-Coal processes, and work is also being carried out at the Japanese Coal Technology Centre. Other Japanese companies with interests in this field include Hitachi, Mitsubishi and Sumitomo. The Ministry of International Trade and Industry (MITI) forecasts that steam-coal imports for all these markets will amount to 53.5mt per annum in 1990 and 80.5mt per annum in 1995.

According to the WOCOL Report, Japan's total coal imports will reach 101–112mtce per annum in 1990 and 132–158mtce per annum in 2000. The main sources will continue to be Australia (46 per cent in 1979), USA (23 per cent) and Canada (18 per cent), with China and South Africa joining the USSR as other important suppliers. The 'develop and import' formula, which has provided Japanese finance and/or equity participation for new mines in Australia, Canada and the USSR, which then export on long-term contracts, will continue to be used.

Growth in coal imports will require large investments in infrastructure. A shortage of deepwater sites for new user facilities will require transhipment terminals of up to 300 000DWT capacity, with onward transit by coaster or slurry pipeline. The NKK steel company is already developing its Fukuyama site as such a terminal, and MITI suggests that three such sites, each handling 40mt per annum, will be in operation by 2000.

The environmental problems of coal use, particularly in Tokyo and other urban centres, were a major factor in its decline as an electricity and industrial fuel. Existing pollution problems in urban areas, together with shortages of sites for both new power stations and ash disposal, and delays in planning procedures, will limit the rate at which coal use can be increased. However, as elsewhere, coal will benefit from public opposition to nuclear power as a base-load alternative.

Japan is highly dependent on imported energy and has therefore been the country most severely affected by the 'oil shocks' of the 1970s. There is no doubt that the formidable powers of the Japanese government will be encouraging much greater coal use and that consumption, especially of steam coal, will rise substantially. However, several factors may undermine the more optimistic forecasts. One is the ability of Japan to maintain a high growth rate in the harsh economic conditions of the 1980s and 1990s. Even if this is achieved, the most likely approach will be a move to high-technology manufacturing and services, with a sloughing off of basic industries, many of which are coal users, to Third World countries. Environmental factors may also constrain coal use, whilst there is some doubt that coal exports will be available in the quantities required or that lead times are sufficient to allow import facilities to be built.[28]

7 Western Europe

Western Europe is the largest coal-importing region and contains several of the world's major producers. Most European countries are planning to expand coal consumption, but output will rise only slightly, leading to a substantial increase in imports.[1]

EEC

Ten European countries – Belgium, Denmark, France, Greece, Ireland, Italy, Luxemburg, Netherlands, United Kingdom and West Germany – are linked in the European Communities (EEC), which has some influence over energy policy. The EEC also incorporates the European Coal and Steel Community, whose responsibility includes coal production.[2] Its ranks will be swelled in the 1980s by Spain, Portugal and perhaps Turkey.

EEC energy policy has been aimed at reducing dependence on imported energy, particularly oil. This has been translated into encouragement for increased coal use, particularly where this is met by community producers, although the practical effect has been relatively unimportant in relation to national governments. During the 1970s, serious consideration was given to an EEC energy policy which would provide financial incentives to encourage the use of domestic energy sources – much of which would be in the form of coal – either by imposing tariffs on energy imports or subsidies to production, on the lines of the existing Common Agricultural Policy.[3] All such policies have foundered on the fact that the countries with the largest energy imports – most countries except the UK and, to a lesser extent, West Germany and the Netherlands – have been unwilling to pay the price which this policy demands. These circumstances are unlikely to change in the 1980s, and it is improbable that intra-EEC flows of coal will rise above present levels. Most of the anticipated increase in coal consumption to the year 2000 is likely to meet by low-cost imports,[4] which are projected by the EEC to reach 140–280mt per annum by 2000,

178

compared to indigenous production of 270–300mt per annum. As at present, much of this production will be subsidised by national governments.

UNITED KINGDOM

Coal has been produced in Britain since the Roman era, and the UK was the world's main source of coal in the eighteenth and nineteenth centuries.[5] Those days are gone, and the industry suffered a calamitous decline in the 1960s and early 1970s, with production falling from 223.6mt in 1957 to 119.9mt in 1978–79.[6] However, recoverable reserves of 45btce and additional resources of 145btce should offer a sound basis for future expansion, although there has been some criticism that these figures are exaggerated.[7] Production stood at 120.9mt in 1982/83, although inland consumption declined to 110.4mt, with 7.1mt of exports.

Geography

The UK contains many coalfields (see Fig. 7.1), some of which are now worked out.[8] Those which remain are:

The *Yorkshire–East Midlands* field produced 59.4mt, over half of NCB deep-mined output, in 1982/83. Most coals are low/medium-sulphur bituminous, which are little disturbed and easy to work compared with other UK deposits. Exploitation has been concentrated in its Western section, where the coal measures reach the surface, but production is now moving to the deeper deposits to the East. This includes the NCB's developments at Selby, where 10mt per annum will be mined from 600mt of reserves by 1990, and eventually, Belvoir. These coal measures extend beneath the North Sea to Continental Europe, but there is no prospect of exploiting these vast deposits until the next century. Most output goes to nearby power stations on the Trent and Ouse river systems.

The *Northumberland–Durham* field produced 12.4mt in 1982/83, of which much was coking quality. However, many pits are exhausting their reserves which, together with a decline in steel output in North-East England, makes continued decline likely.

The *South Wales* coalfield is extremely geologically disturbed, with

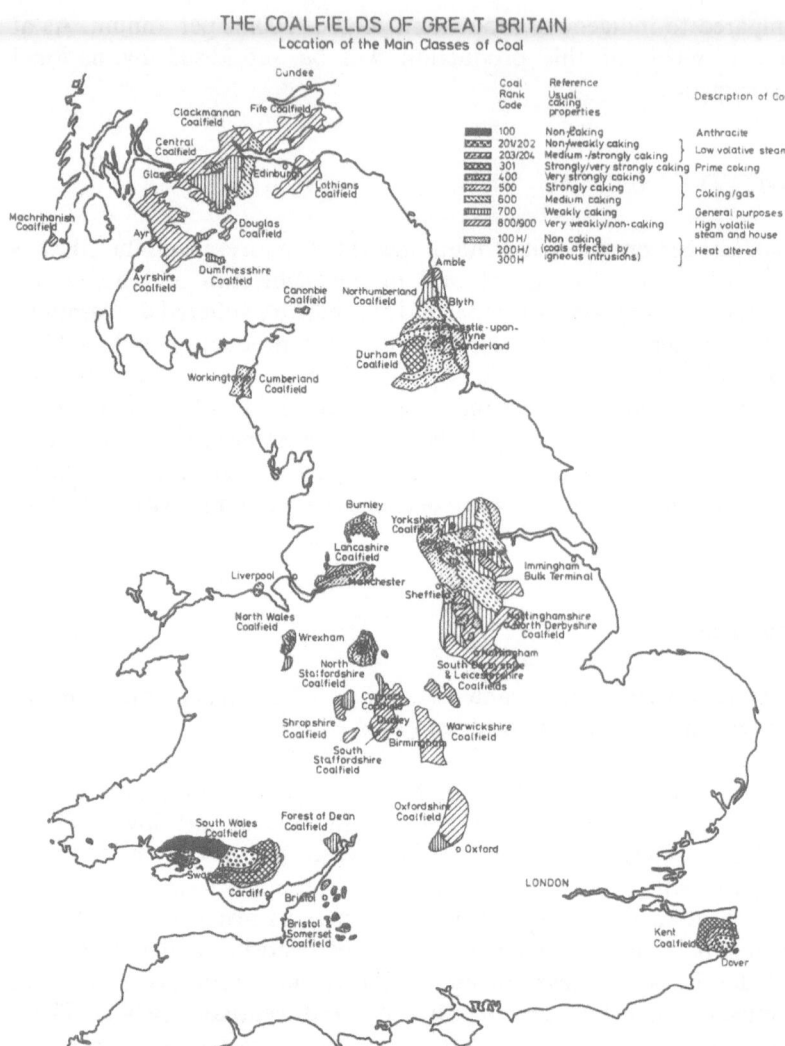

FIGURE 7.1 The coalfields of Great Britain – location of the main classes of coal

SOURCE National Coal Board.

most coals present as high-ranking bituminous or anthracite. Output in 1982/83 was 6.9mt, a decline on the previous year, and produced a £113 million loss. Exhaustion of reserves, the collapse of the South Wales steel industry and high mining costs make it likely that output will continue to fall, despite some new investment.

The *Scottish* coalfields are also geologically disturbed and have been subject to extensive igneous activity. Production was 6.6mt in 1982/83, and future markets will be limited by decline in the steel industry and construction of base-load nuclear power stations in the region.

Other working coalfields are located in Lancashire, Kent, South Derbyshire–Leicestershire, North–Central Staffordshire and Warwickshire.

NCB open-cast (surface) mines produced 14.7mt from all coalfields in 1982/83. High profits from this source helped offset losses in many of the other areas.

During the twentieth century an increasing percentage of UK coal output has come from the central coalfields of East Midlands–Yorkshire, Staffordshire and Warwickshire. This has been due to adverse geological conditions and exhaustion of reserves in 'peripheral' areas, which have been mined for many centuries. In South Wales, for example, Output per Manshift (OMS) in 1982/83 was 1.47 tonnes, at a production cost of £61 per tonne, compared to the national average of 2.44 tonnes at £41 ($60) per tonne. The problem has been compounded by the central coalfield's accessibility to markets. As a result, between 1960 and 1977, saleable coal output fell by 62 per cent in the North–East Division, 57 per cent in the South–Western, 52 per cent in the West and 50 per cent in Scotland, compared to only 25–30 per cent in Yorkshire and the Midlands. Job losses and colliery closures also fell disproportionately in the peripheral areas with the result that the central fields accounted for 53 per cent of the NCB workforce and 62 per cent of the output by value in 1977, compared to 39 per cent and 48 per cent respectively in 1960.[9]

This trend has continued. Most new investment is being channelled into the central areas, and the new coalfields which are likely to be opened in the next 10–20 years – Selby, Belvoir and Warwickshire – all fall within it. The prospects of the peripheral fields are also blighted by adverse market conditions, particularly decline of the steel industry and increased coking-coal imports. As a result they were disproportionately affected by the NCB's production cutbacks of 1981, although these were later revoked. In the long run, their output will continue to fall.

Organisation

Almost all coal-mining operations in the UK are carried out by the state-owned National Coal Board (NCB) which, since its formation in 1946, has been the largest coal producer in the Western world. In 1982/83 it had a turnover of £4.9 billion ($7 billion) and employed 266 000 people, making it one of the largest commercial organisations in the UK.[10] Underground mining, which accounts for almost 90 per cent of output, is decentralised to twelve mining areas, each controlling a number of collieries. Responsibility for surface mining lies with an Opencast Executive, which sub-contracts the physical operation of sites. Two subsidiaries, NCB (Coal Products) Ltd. and NCB (Ancillaries) Ltd., have interests in smokeless-fuel manufacture and coal-based chemicals, and coal distribution, mining consultancy and computer services. The NCB also operates one of the largest coal research programmes in the world at its Mining Research and Development Establishment (MRDE) and Coal Research Establishment (CRE), whilst International Energy Agency (IEA) coal research is carried out in the UK, under NCB auspices. The NCB is also allied with equipment manufacturers and other private companies in British Coal International Ltd., which provides technical and consultancy services to overseas markets.[11]

During the 1960s and early 1970s the NCB, like other European coal organisations, was badly hit by the industry's decline. Morale suffered at all levels, with employees being transferred from one pit to another to escape closures, or choosing to leave the industry to greener – or less black – pastures elsewhere. This situation changed after the 1973 energy crisis, but it has taken some years for a new sense of dynamism to suffuse through the industry, and for the NCB to obtain the human and financial resources to implement its optimistic plans for the future. The basis of the NCB's present strategy is its 'Plan for Coal', agreed by a tripartite group of management, unions and government in 1974.[12] This forecast that markets for coal in 1985 would be 125–150mt per annum, although the estimate for that year was later reduced to 121mt. To meet this demand, the Plan recommended the construction of 42mt per annum of new capacity to replace depleted or uneconomic mines, of which 20mt per annum was to be from totally new schemes. In addition, it also envisaged that open-cast production would rise from 10mt per annum in 1973/74 to 15mt per annum in 1985. Finally, it proposed increased expenditure on new mining and coal-use technologies, and an expanded exploration programme to discover further reserves.

Considerable progress has been made in the implementation of 'Plan for Coal'. Large capital investments in Selby and other new or reconstructed mines have taken place, and new mining equipment developed. In particular, the Heavy Duty Mechanisation (HDM) programme has introduced more ruggedly-built face equipment to improve reliability and safety, and reduce maintenance costs, while the Advanced Technology Mining (ATM) programme has introduced a greater degree of sophistication in other equipment, particularly powered supports.[13] Daily output in the 6.4 per cent of faces using HDM and ATM technology in 1982/83 was 1 413t, more than twice that from conventional longwall faces. Work is also proceeding on the MINOS control system, which will bring the semi-automated mine into being. As a result of these, and other factors, output and productivity increased during the late 1970s and early 1980s.

However, considerable difficulties face the NCB and its continued implementation of 'Plan for Coal' in the 1980s. One is the escalating cost of its investment programme, which rose from £1.4 billion in 1974 to £3.2 billion ($5 billion) in 1980 – a 43 per cent increase in real terms. This, together with delays in commissioning new revenue-earning capacity and the need to finance large pithead stocks of coal, has put the NCB heavily into debt. In 1982/83 this was over £3 billion, requiring £366 million interest payments. The NCB's deficit rose to £485 million, and the Board was 'technically insolvent'.

The 1980 Coal Industry Act eased some of the burden of debt financing and gave some additional social grants, but set unrealistic financial targets for the NCB.[14] The situation was worsened by a decline of sales due to recession and the high value of sterling, so that in 1981 the NCB was forced to introduce plans to reduce its production capacity and declare workers redundant. Under threat of strike action by unions, the government agreed to give financial support of approximately £400 million to the NCB and to introduce restrictions on imports. This led to the withdrawal of the closure plans and stabilised the situation in the short–medium term, although further difficulties in this area may occur in the future.

The main union involved in the campaign against the NCB's production cutbacks has been the National Union of Mineworkers (NUM).[15] This organises most NCB employees, with the exception of some white-collar and skilled staff. It is unusually decentralised, with its constituent geographical and occupational branches having considerable autonomy. Political factions are also well organised and draw their support from particular coalfields – thus, Yorkshire, South Wales and

Scotland are traditionally militant while the Midlands is usually moderate.

Before 1946, the relationship between mineowners and workers was one of extreme bitterness, and nationalisation was primarily due to the unwillingness of miners to accept the free enterprise system. This positive support of the workforce for its existence has given the NCB a considerable advantage which, despite considerable disillusion, it has retained to the present day. This is reflected in its system of worker consultation which, although criticised within the industry, is one of the more advanced in the UK. At national level, unions, management and government meet in tripartite sessions to determine future policy for the industry – as with Plan for Coal. This is paralleled by a Joint Policy Advisory Committee of NCB board members and union officials, which has regular discussion of current results and future plans. Beneath these are consultative committees at area and colliery level. Partially as a result, coal mining was responsible for only 13 per cent of working days lost in industrial disputes in the period 1945–80, compared to 50 per cent before it.

The present character of coal industrial relations has been shaped by mining's decline in the 1960s and early 1970s. Employment and the number of collieries fell dramatically, and – despite accepting this rationalisation and the mechanisation which accompanied it – miners fell from being third in the national earnings table in 1960 to twelfth in 1970. As one study commented: 'This rundown, which was brought about with the co-operation of the miners and their union, is without parallel in British industry in terms of the economic and social costs it has inevitably entailed for the mining community as a whole'.[16]

The frustration of miners during this period were expressed in the sanctions and strikes of the winter of 1971–2, which the unions won against (Conservative) government opposition. A second strike, in 1973–4, sought – in the wake of the oil shortages caused by the Yom Kippur War – to restore mineworkers to their old status and position, and was again opposed by the government. A general election was called on the issue, which the Conservative government lost, allowing the pro-NUM Labour Party to return to power. Since then, the question of coal has been a subject of great political controversy.

The 1974 strike established miners as one of the most secure and highly-paid sections of the British labour force, and industrial relations have since been good. Output per manshift (OMS), which increased markedly from 1.25 tonnes in 1960 to 2.21 tonnes in 1970, stagnated for most of the 1970s, but then recovered to stand at 2.44 tonnes in

1983. The end of the 1970s also saw a reduction in absenteeism, and a gradual fall in the high average age in the workforce as recruitment improved. The NCB has attributed these improvements to the introduction in the late 1970s of a new incentive payments scheme. Although this faced initial union opposition, the NUM seems to have tacitly accepted a high-wage, high-productivity industry as its goal. In line with this aim, the NUM has also succeeded in negotiating fringe benefits, such as early retirement, which are often ahead of the rest of the world.

As the events of 1981 showed, these successes may be overshadowed by the financial problems of the NCB. Although mining unions forced the NCB to revoke its plans, the financial cost to both the Board and the government of maintaining old capacity is considerable, and further closure plans may be formulated in the future. Even if this is not the case, geological problems may force some mines to cease production, although in certain circumstances this too may be opposed by unions. This tension caused by the NCB's desire to maximise the share of modern, high-efficient pits in its productive capacity is likely to cause problems for much of the 1980s and may seriously jeopardise the development of the industry.

Markets

In 1982/83, the electricity generating industry used 80.8mt of coal – compared to 89.1mt in 1979/80 – and this market will remain the key to the NCB's fortunes until the end of the century. Since the early 1970s, coal-burn in UK power stations has been encouraged and subsidised by government. In 1977 the main UK utility, the Central Electricity Generating Board (CEGB), committed itself to a coal-burn of 75mt per annum, although this was dependent upon the market price of coal remaining in line with that of oil. During 1978/79, a temporary Assisted Coal Burn scheme was introduced to help the NCB weather an 'increasingly difficult marketing situation', and to further the government's strategy of reducing oil consumption. Under this, the CEGB agreed to increase coal-burn by 2.5mt per annum and to reduce imports 1.75mt per annum. The extra cost – about £19 million – was given as a subsidy to the NCB, who used it to reduce the price of coal. A separate agreement, involving subsidies of £1.5 million, was reached with the South of Scotland Electricity Board (SSEB) to increase coal-burn.

The 1980 Coal Industry Act, and the medium-term strategy being followed by UK utilities, creates some uncertainty as to the future power-station market available to the NCB. Both CEGB, and government, policy is based on a major nuclear construction programme to provide much of future base-load demand, for which it is thought cheaper than other fuels. In the short–medium term, coal will therefore be competing with oil for base- and peak-loads, with the outcome partially depending upon the disparity between coal and oil prices. The CEGB has estimated that by 1984/85 its coal demand will be between 75–85mt per annum, of which imports would comprise up to 10mt per annum, compared to 4.5mt in 1980.[17] Although the CEGB has stated that the NCB may have difficulty in reaching these requirements, adverse circumstances – such as reduced rates of increase in electricity demand, narrowing of coal–oil price differentials or a decision to import more coal – could make permanent the reduction in NCB sales to this market which occurred during the 1980s' recession.

In the medium–long term, the problem will be compounded by the obsolescence of much existing capacity. Only one new coal-fired power station has been ordered since the 1960s (Drax B) and no further orders are planned, although some oil-fired stations may be converted to coal-burning. At the same time, much of the 30–40 000MWe capacity which will be closed by 2000 will be coal-fired, as will some of the stations which are kept in use but demoted in the CEGB's merit order. If new coal-fired capacity is not ordered in the 1980s – and the CEGB maintains that electricity is produced more cheaply from nuclear power than coal – utility coal-burn can be expected to decline in the 1990s. A government study has estimated that it will be 78mt per annum in 2000 (compared to 94mt per annum in 1990) , and other official estimates have seen it declining to 30mt per annum in the early twenty-first century.[18] The WOCOL Report also projects that this market will require 70–7mtce per annum in 2000.[19]

NCB sales of coking coal have fallen as a result of the problems of the UK steel industry. In particular, the state-owned British Steel Corporation (BSC) suffered huge losses in the 1970s, and drastically reduced its capacity during the period 1978–80. In 1982/83, this market consumed 7.6mt of coal – compared to 29mt in 1960 and 10.7mt in 1979/80. Intense foreign competition makes it unlikely that UK steel output will reach its former levels, and BSC will be tempted to cut costs by using more imports. The WOCOL projection of 16–21mtce per annum for this market in 2000 is therefore highly optimistic.

The NCB has high hopes of the UK industrial coal market, which consumed 8.8mt in 1982/83 – compared to 10.7mt in 1979/80 – and is seen as 'the most important growth market for coal for the rest of the century', especially when new technologies such as fluidised-bed combustion are introduced.[20] However, demand in this market, as in others, dropped during the early 1980s. The WOCOL Report estimates demand in 2000 at 21–6mtce per annum, although an increase of this size will depend upon the performance of manufacturing industry in intervening decades, and favourable trends in gas prices and availability. The latter factor will also affect the performance of coal in the residential and commercial sector, which consumed 8.0mt in 1981/82. The WOCOL Report projects that this will reach 15–21mtce per annum by 2000. Coal conversion is also expected to provide a small market by 2000, although this will not become substantial until the next century.

Coal exports reached 7.1mt in 1982/83 – compared to 2.5mt in 1979/80 – with France, West Germany, Scandinavia and Israel providing the main markets. A proportion of this – the actual amount being the subject of controversy – is sold at or below cost price to obviate the rising stocks and reduced cash flow caused by decline in UK cash consumption. The NCB hopes to further increase exports in the 1980s but this will prove difficult in the face of cheap imports from non-European sources. Coal imports have been restricted, and these fell to 3.4mt in 1982/83. Both the CEGB and BSC are interested in increasing the total, although this may be prevented for political reasons. The effects of the recession, and the agreement reached between the government and miners unions in early 1981 will produce a decline in imports for several years, but these may rise again in the middle of the decade.

Environment and Infrastructure

The environmental and social impacts of coal production in the UK are highly controversial, due to the high value placed on rural land in a densely populated country and a long history of adverse environmental consequences from past mining.[21] The relics of this, in the form of waste tips and subsidence, still blight the landscape in many parts of the country. Although operating practices have considerably improved under the NCB, concern about waste disposal, air and water pollution and subsidence caused bitter opposition to its plans for new deep mines and expansion of open-cast (surface) mining at many sites. This opposition will become more, rather than less, intense and will, at best, create extra costs for the NCB as stringent environmental conditions are attached to mining approvals and, at worst, completely prevent the opening of some new capacity.

In recent times, coal use has been less controversial, largely because of the dramatic improvements brought about by the 1956 Clean Air Act and subsequent modifications. Although concern over the trace-element emissions from, and cooling-water needs of, power stations has been expressed, the question of sulphur-dioxide emissions is the most significant issue. The CEGB controls these by dispersal from tall stacks, which Scandinavian countries claim leads to major problems of acid rain (also derived from Continental emissions) over their territories. A combination of external and domestic pressures may lead to the introduction of flue-gas scrubbing before the end of the century, raising the price of coal-derived electricity with possible effects on NCB sales. The environmental problems of existing coke ovens and smokeless-fuel plants may also affect the construction of coal-conversion complexes in the future.

The UK has a sophisticated transport system, and this is unlikely to constrain future coal production or use. Most coal will continue to be moved by 'merry-go-round' unit trains, although an increased percentage will make use of upgraded canals – such as the South Yorkshire Navigation – or coastal ships. An increase in coal imports or exports will require new port facilities, probably on the Humber, Mersey or Severn estuaries, although some use could be made of the existing coking-coal terminals at steel ports.

Coal and Energy Policy

The United Kingdom is a paradox in the world coal industry. Although the country has a large reserve base, sophisticated technology, highly-skilled labour and developed markets, it is one of the few countries where the coal outlook is bleak in the short–medium term. Four factors explain this situation. One is the energy riches of the UK, which has large reserves of oil and, more important, gas, which limit coal's penetration of new markets. High production costs and other, non-economic, reasons have also led to government and utility support of nuclear power, reflected in the new capacity which will be commissioned in the 1980s and 1990s. The third is the government's setting of unrealistic financial targets for the NCB, which has affected its price competitiveness and exacerbated its industrial relations problems. Although the situation was alleviated by the 1981 agreement between the NCB, unions and government, similar problems may occur in the future. The fourth, and most important, factor is the state of the UK economy, which has a low rate of growth, is affected by an exchange rate buoyed by a 'petrocurrency' status, and suffered disproportionately in the world recession of the early 1980s. As a result, coal sales, and production, are unlikely to significantly expand for much of the 1980s, and may even decline, as in the recession of 1980–1 when UK coal demand fell by 6mt. Any expansion in industrial demand – which will be affected by relative shrinkage of the manufacturing sector in the UK economy – seems likely to be offset by stagnation in electricity and metallurgical markets, where prices will be controlled by the availability of low-cost imports.

Several observers have argued that the NCB may be in trouble until the 1990s, with one suggesting a market of only 98–105mt in 1990 and 107–116mt in 2000, and another 70–110mt by 2000.[22] This seems unlikely, as the current modernisation programme and phasing out of old capacity will improve the NCB's position – new mines at Belvoir and Selby will have productivity several times higher than existing pits, whilst the MINOS electronic control system will also produce great benefits – and new markets can be expected to develop in the 1990s, in a context of world energy shortage. In addition, political considerations make it unlikely that governments would allow another large-scale contraction of the coal industry beyond the levels set in 1981, although it is possible that there will be no significant increase. Government forecasts suggest an NCB output of 127–138mt in 1990 and 137–155mt for 2000, compared to a WOCOL Report projection of 133–162mtce per annum by 2000, with the circumstances of the 1980s suggesting that the lower end of the these ranges will be the most likely outcome, and may even represent an upper limit.[23]

WEST GERMANY

The decline of the West German coal industry in the 1960s and early 1970s was one of the most precipitous and traumatic of any nation. Between 1957 and 1977 the number of collieries fell from 173 to 42, hard-coal output from 150 to 86mtce per annum and employment from 600 000 to 180 000. Since 1973, West Germany's dependence on imported oil and rising energy prices have produced, as elsewhere, a turnround in the industry's fortunes. Hard-coal production was 88.5mt per annum in 1981 and will increase in the 1980s, although not by a sufficient amount to meet all the projected increases in demand. In 1981, lignite production of 130.7mt (38mtce) was added to this total. Recoverable reserves stand at 34.6btce, with additional resources of 186.3btce, ensuring the industry's survival well into the next century.[24]

Geography

The Ruhr Basin has always been the mainstay of West German production, and now accounts for 80 per cent of national hard-coal output, of which 90 per cent is produced by Ruhrkohle AG. Deep mining is the norm, with slightly more being sold for coking than electricity generation. However, these quality advantages are partially offset by difficult geological conditions and variability within seams, so that coal needs considerable sorting above ground.

The Ruhr mining zone has shrunk by 55 per cent to about 850 square kilometres in 1979, and is now fragmented into individual sub-zones. Production has gradually concentrated in the north of the field, with the area to the south of the Emscher river accounting for only 10 per cent of output in 1979, compared to 50 per cent in 1956. Remaining virgin coal measures in the Ruhr field are also to the north of present mining zones.[25] Further reserves are available at greater depths (below 1 000 m) but exploitation is hampered by their relatively low coal content – 4–6 million tonnes per square kilometre (compared to 10–12 in existing areas) – and the varying thickness and quality of the seams. In addition, there is some possibility of opening up abandoned reserves by reconstruction of existing pits, as in the Gelsenkirchen-Harten area.

The Saar Basin has suffered severe rationalisation, but its present six pits are the most highly mechanised and productive in Europe. Its high-volatile bituminous coal accounts for 11 per cent of West German output, most of which is used for electricity generation. Substantial virgin reserves exist to the north, south and east of the present mining zone, whilst an extra 4–5mt per annum of production is thought to be possible from the reserves of past mines, which were closed in such a way as to render them re-usable in the future. Original targets of 17mt per annum output by 1995 have been trimmed, but production is expected to reach 11.2mt per annum by 1985. Under the 1957 Saar Treaty, one third of the field's output can be sold in Lorraine, whilst parts of the field are allowed to be mined by French pits across the border. The impending expiry of this treaty, together with the economic problems of Lorraine, may hamper sales possibilities in the future.

The Aachen Basin is an extension of the Franco–Belgian fields, with the same complex geology. It contains three collieries, which produced 6 per cent of West German output in 1979, mainly in the form of anthracite and coking coal. Untapped reserves, plus the possibility of tapping unmined seams in closed collieries, should guarantee production on the present scale until the next century.

Ibbenbühren is an isolated mine in North Germany, reprieved from closure from social and employment reasons. It now accounts for 3 per cent of national output, and its proven reserves should extend production until early in the next century, although new ones are unlikely to be discovered. At 1443m, it is the deepest mine in the world.[26]

The Rhenish lignite field, to the West of Cologne, produces most (90 per cent) of West German lignite output, by surface mining seams up to 100m in thickness.[27] Although older mines are being exhausted, production in the 1980s will be boosted by the Hambach mine, whose output should reach 40–50mt per annum (15–17mtce per annum) by 1990, at an investment cost of DM5 billion. Although the huge scale of mining operations in a densely-populated area creates great environmental and social problems – including the movement of whole villages – the sole producer, Rheinische Braunkohlenwerke AG, has so far resolved these with local inhabitants.[28]

Similar lignite deposits are tapped by other companies in the vicinity of the East German and Czechoslovakian borders.

Organisation

Responsibility for the coal industry is divided between the Economics and Technology Ministries. In fact, West Germany has a uniquely-organised coal market, based on a great amount of government intervention and close cooperation between coal producers and consumers.[29] Government aid to the industry takes seven forms: (a) *Investment Aid* in the form of subsidies to new projects for mines, coking and briquetting plants and pithead power stations. (b) *Underground worker subsidy* amounting to DM5 per shift per man as an incentive to mining companies to retain miners. (c) *National coal reserve* – since 1974 the government has been purchasing for a 10 million tonnes stockpile, thus underpinning coal prices. (d) *Innovation grants* to aid the introduction of new machinery and techniques. (e) *Special depreciation subsidies* to encourage investment in underground production. (f) *Coking coal subsidies* to maintain indigenous coke-making in the face of foreign competition. In addition to these categories, government funds contribute to coal research and development expenditure, and social expenditure on miners and their families. An example of how powerful these incentives can be in affecting the shape of the industry is given by the Ibbenbühren colliery in North Germany. In 1974, the workforce had declined to 4 000 and its reserves were almost exhausted, so that the pit was on the point of closure. Government help, including a DM90 million loan, persuaded the owners, Preussag, to explore further reserves and, when discovered, to exploit them at a profit.

In addition to direct government support to coal producers, large subsidies are also given to public utilities which burn guaranteed amounts of coal (see below). Since the mid-1970s this has been in the form of a 'Kohlenpfennig', a levy imposed on electricity consumers to make up the difference between the cost of burning German coal and imported oil. Initially, this raised large sums of money, but its burden declined after 1979 as oil prices rose.

The net cost of all these measures in 1978 was DM4.1 billion (£1 billion), amounting to DM45.6 (£11.90) per tonne of coal sold. Since then, the level of subsidy has dropped as coal became more competitive with oil. In addition to these arrangements, the West German industry has also received aid from the European Coal and Steel Community.

All West German hard-coal production comes from underground mines, usually equipped with mechanised longwall faces.[30] High levels of investment have made the industry highly productive, with its 182 000 workers achieving a face output per man-shift of up to 20 tonnes in 1979. Despite this, poor geological conditions and the high value of the Deutsche Mark make production costs extremely high–DM163 ($70) per tonne in 1978, compared to spot prices for imported steam coal of $33–36 in the same year – and the industry could not survive without the subsidies detailed above. It is estimated that, on an assumption of 4 per cent annual growth in miners' real wages, the average cost of domestically-produced coal will grow, in 1978 currencies, to DM173 in 1985 and DM227 in the year 2000 – almost certainly a greater increase than in other coal-producing nations. To counteract this tendency, great emphasis will be placed on further automation and reconstruction of existing collieries. According to one estimate, use of existing in-site facilities, such as winding shafts, can reduce lead times for new capacity from twelve to eight years, and capital costs from 5–600DM/tonne of annual production to 350–400DM. Total capital needs for the projected expansion of hard-coal production to 95mt per annum in 1990 and 100mt per annum by the late 1990s are estimated at DM8 billion for new capacity and DM3 billion per annum for the maintenance of existing mines.[31]

Labour relations in the coal industry have generally been good, despite the fact that coal miners are not at the top of the industrial earnings table, as is the case in some other European countries. One factor has been the application of co-determination laws – giving workers a say in management – to the industry. This has also contributed to the marked improvement in health and safety which has occurred in recent decades. Between 1958 and 1978, cases of silicosis amongst miners declined by 80 per cent, whilst their average life expectancy rose from 54 to 73 years.[32]

The West German coal scene is dominated by Ruhrkohle AG, formed in 1968 from a government-backed merger of most of the Ruhr's coal mines.[33] Its twenty-nine mines now account for three-quarters of national output, and its twenty-one coking plants are amongst the most efficient in the world – to the extent of processing US coal for re-export to that country. A coal-trading subsidiary, Ruhrkohle-Verkaufsgesellschaft, sells approximately 14mt per annum of mainly coking coal and coke to thirty-nine countries, giving it about 10 per cent of total world coal trade. The company is undertaking research into coal gasification, hydraulic mining and coal-combustion techniques, and has holdings in other energy sectors, including a controlling stake in the STEAG power company and a minority stake in gas and chemical companies. Overseas interests are grouped under Ruhrkohle International, and include a 50 per cent share in the Saxon mine of British Columbia and a 10 per cent stake in the German Creek development in Australia, both of which should produce 3–4mt per annum by the mid-1980s. By the year 2000, the company expects to be handling more than 10mt per annum of overseas coal. To facilitate blending of these coals with German production, the company also has a holding in a new Rotterdam coal port. The other major German producers are Saarbergwerke SA, formed in a similar manner to Ruhrkohle in order to rationalise the Saar mines, and Rheinische Braunkohlenwerke, owned by a consortia of utilities. Scientific and technical work is carried out by a joint agency, the Steinrohlenbergbauverein.

Markets

Most German coal output – 110mt of lignite and 40.4mt of hard coal – is used for power generation, providing 58 per cent of total sales. Coal-fired capacity is planned to increase from 30 000MWe in 1980 to

38 000MWe in 1990 and 55 000MWe by 2000. By comparison, the WOCOL Report sees a necessity for 47–49GWe for hard-coal capacity and 18GWe of lignite capacity by the year 2000. This would produce a demand for hard coal of 60–66mtce per annum and for lignite of 39–40mtce per annum and account for 38–44 per cent of installed generating capacity.

Penetration of the electricity market has been aided by government intervention, both in the form of the 'Kohlenpfenning', and a guarantee by the utilities – given under official pressure – that they would burn at least 33mt per annum of domestic coal between 1978 and 1987. This provided the necessary security for new investment programmes to be begun by the coal producers.

These measures were extended in 1980 by the so-called 'agreement of the century' between government, producers and utilities.[34] Under this scheme, the utilities increased their guarantees of domestic coal purchases to 40mt per annum by 1985, 45mt per annum in 1990 and 45–50mt per annum by 1995, which covers half of future coal production. The legislation which brings the agreement into operation provides the utilities with government subsidies based on the difference in price between domestic and imported coal. In addition, they will continue to receive the 'Kohlenpfennig', although this is expected to decline further in value as coal–oil price disparities are reduced.

By guaranteeing larger markets for domestic coal producers, the agreement also removed opposition to coal imports, which will increase substantially in the next two decades. Under its terms, utilities will be allowed to burn a tonne of non-EEC coal for every 2 tonnes of German coal burnt above 33 million tonnes per annum between 1981 and 1987, and for every tonne of German coal burnt after 1987. The formula should mean that utility imports average 4mt per annum until 1985, 8mt per annum from then until 1990, and 12mt per annum thereafter. This is in addition to existing agreements allowing coastal utilities to import 5mt per annum, which has been extended until 1985 and will almost certainly remain in existence thereafter.

The iron and steel industry consumes a further quarter – 20.3mtce in 1981 – of West German hard-coal production. Despite the problems of this sector during the 1970s, demand – based on its excellent coking characteristics – is expected to remain at a similar level until the year 2000. However, costs may be reduced by the blending of German and cheap imported coals in coke ovens.

Present industrial and residential/commercial markets amount to 7.8mtce of hard coal (approximately 10 per cent of production) and

3mtce per annum of lignite. Industrial use of coal is expected to increase to 8–12mtce per annum by the year 2000, when residential/commercial consumption will have declined slightly to 8–10mtce per annum. As government will be reluctant to subsidise private industry, and as the increase will be direct substitution for other fuels, most of this market will be supplied by cheap imports. Under the 1980 agreement, the then import ceiling of 1mt per annum was lifted to 4mt per annum until 1986, 8mt per annum until 1991 and 12mt per annum from then until 1995, although it is unlikely that these levels will be achieved.

Despite its high production costs 17.2mt of German coal was exported in 1981 – mainly to France and Belgium – as well as considerable quantities of coke.[35] Half of these exports go to the iron and steel industries of other EEC countries. As more coal is diverted to domestic consumption, exports will not rise above 15–20mtce per annum until 2000. As with domestic consumption, coke exports will increasingly be made from blends of German and imported coals.

West Germany is currently the world leader in coal-conversion technology, with the three gasification technologies presently in existence – Lurgi, Koppers-Totszek, Winkler – all belonging to German companies. An extensive R & D programme is presently developing improved versions of these processes, as well as work on liquefaction.[36] Under a government-supported plan announced in 1980, up to DM13 billion was to be invested in fourteen gasification and liquefaction plants in the Ruhr and Saar, although this was later cut back. This would provide markets for 12mt per annum of hard coal and 10mt per annum of lignite, although under the 1980 agreement up to 5mt per annum may be obtained from coal imports. According to the WOCOL Report, demand from this sector by the year 2000 could reach 10–25mtce per annum. Many observers believe that the high production costs of West German hard coal make economic conversion unlikely until the next century, although it is estimated that gasification of domestic lignite would be profitable at 1980 oil prices. In practice, the West German coal industry would be hard pressed to provide this amount of coal, as well as increased supplies to electricity and other markets, so that a higher percentage of this sector would be supplied by imports. The economic and budgetary difficulties of the early 1980s have also made it likely that expenditure in this area will be restricted for some years.

Until the signing of the 1980 agreement, coal imports had been tightly regulated by the government, but should now rise rapidly. According to the WOCOL Report, total imports will grow from 11.3mt in 1981 to 20–40mtce per annum in 2000, with possible delays to the nuclear programme making the upper figure the most likely. The main sources are expected to be Australia, the USA and South Africa.

Environment and Infrastructure

West Germany has a sophisticated and efficient transport system and no difficulties seem likely to occur in the future. Environmental problems may present more of a constraint on future production and use, particularly in view of the political importance of the 'Green' parties and other ecological groups. Their capacity to delay coal-related development was demonstrated by the Vöerde power plant in the Ruhr, whose construction was halted for three years on environmental grounds. For coal mining, the environmental effects of surface lignite production, which are concentrated near to the large cities of Köln, Dusseldorf and Aachen, may become more controversial, whilst demands may grow for a tightening of the Federal Emission Control and Water Administration Acts to reduce pollution from coal use – perhaps including a requirement for flue-gas desulphurisation technology to be installed. Finally, domestic pressure may be strengthened by external pressures, as from Scandinavian countries, which complain of a high incidence of 'acid rain' as a result of German (and other EEC) coal use. EEC environmental measures may also have an impact in this area.

The licensing of new generating capacity may also be delayed by environmental objectors, as has been the case in the 1970s. In the short – medium term, however, coal may be a beneficiary from a more intense opposition to nuclear power, producing a view that it is the 'better of the two evils'.

Coal and Energy Policy

With over half of its energy needs presently provided by oil imports, West Germany is relying on coal and nuclear power to underpin its future economic prosperity.[37] Accordingly, its high-cost domestic industry has been greatly subsidised and this situation is likely to continue in the future. The result should be a modest expansion of domestic production to 100–110mtpa by 2000, with investment in new capacity encouraged by the existence of guaranteed future markets. At the same time, the 1980 utility–industry agreement will ensure that some advantage can be taken of cheap imports, without damaging domestic interests.

The controversial nature of energy policy, and environmental protection, in West Germany will ensure that the plans for coal will be subject to searching scrutiny. On balance, they are likely to be accepted by many as a less objectionable alternative to nuclear power. Encouragement will also be derived from the mining and coal-conversion equipment industry, whose importance seems likely to increase in coming decades. As a result, projections such as those of the WOCOL Report – consumption of 150–175mtce per annum in the year 2000, of which imports would provide 20–40mtce per annum – can be considered more reliable than in the case of some other countries.

BELGIUM

Although the Belgian coal industry is one of the oldest in the world, its production declined to 6.1mt in 1981, with a further 10mt of imports.[38] Recoverable reserves now stand at 440mtce, with additional resources of 2.6btce. Over 90 per cent of output is from the Campine (Limburg) coalfield in North-East Belgium, with the rest coming from the Southern Basin, an extension of the French Nord-Pas de Calais field which has suffered a similarly large loss of production as a result of adverse geological conditions. The government's policy is to stabilise output at 6–7mt per annum for the rest of the century, which will require substantial investment. Belgium has also conducted extensive research on underground coal gasification, with a large experimental station at Hensies, and this may allow presently uneconomic deposits to be exploited in the medium–long term.

Coal consumption is concentrated on the iron and steel industry and electricity generation. Although the former market will continue to be hit by cutbacks in capacity and foreign competition, this should be more than balanced by conversion of oil-fired power stations to coal, construction of new coal-fired capacity and increased industrial use.

DENMARK

Denmark has no coal resources, although bituminous deposits occur in Greenland, its autonomous dependency.[39] Despite this, over 60 per cent of electrical generating capacity is coal-fired, and this is expected to reach 85–90 per cent by the 1990s. However, environmental regulations in Denmark are mild and increased coal use may be restrained by new measures, particularly over the sulphur-dioxide emissions which contribute to the problems of acid rain experienced by other Scandinavian countries. Other, minor, users of coal are industrial and district heating plants, but future consumption by these is unlikely to rise by significant amounts.

The WOCOL Report projects coal consumption in 2000 of 9.4–20.9mtce per annum, compared to 10.9mt in 1981. Several developments initiated after 1979 suggest that the higher figures are likely to be achieved. Until 1980, most Danish coal imports came from Poland, but disruption of supplies from that source led to a hasty, and costly, diversification of suppliers. This process will continue in the future particularly through a contract signed with Colombia for 2mt per annum of imports by 1986. Although deep-water facilities already exist in South Jutland and West Zealand, new ports will be needed, perhaps in the form of a 300 000DWT transhipment terminal serving all the Scandinavian countries.

FRANCE

Coal fell from 70 per cent of French primary needs in 1950 to only 18 per cent in 1978, with a corresponding decline in production.[40] In 1981 this amounted to 23.2mt, dwarfed by imports of 27.4mt. Recoverable reserves are 574mtce with additional resources of 227mtce, and output will continue to fall to approximately 12mt per annum by 1990, despite a rise in coal consumption.

The *Lorraine* coalfield, an extension of the West German Saar Basin, produced 9.6mt in 1979 and contains two-thirds of French recoverable reserves. As mining and geological conditions are generally favourable, French output will be increasingly concentrated on it in the future. Although Lorraine coal has poor coking qualities, large quantities are blended for metallurgical uses.

The *Nord-Pas de Calais* coalfield has been productive in the past, but exploitation is increasingly hampered by adverse conditions, with thin seams and a highly-faulted and folded geology. Although new investments in the 1970s have extended its life, its 1979 output of 5.4mt is expected to gradually fall in the remainder of the century. Until then its varied bituminous coals, many of coking quality, will continue to be consumed in local metallurgical industries and for power generation. The *Midi-central* fields are a number of small basins in central France, many of which ceased mining during the 1960s and 1970s. Output in 1979 was 5.5mt, mainly for local metallurgical industries, and further decline is inevitable.

Other coal deposits which are exploited include anthracite at La Mure, in the Northern Alps, and lignite in Provence.

The industry itself is dominated by a nationalised body, Charbonnages de France (CdF), which also has investments in Australian and American coal properties, and extensive chemical interests. Its subsidiary, CdF International, aims to produce 10mt per annum in overseas mines by 1990. In 1979, these other interests balanced coal losses to give a profit of Fr60 million, the first in six years. France's substantial imports are handled by an associated body, the Association Technique de l'Importation Charbonniers (ATIC). The ELF and CFP (Total) oil companies also have coal interests in the USA and elsewhere.

Although high investments in the 1970s have modernised most collieries, adverse geological conditions restrict productivity and output in the Midi and Nord fields. Although these are partially offset by the efficiency of the Lorraine mines, overall productivity levels are well below those of West Germany and the UK. However, the labour force fell by 5 200 to 64 000 during 1979, and the continuing closure of less efficient capacity will continue during the 1980s.

Since the 1973 oil crisis, Electricite de France (EdF) has had a programme of substituting coal for oil wherever possible, and electricity generation used 17.6mt of coal in 1979, divided between domestic and imported sources. The WOCOL Report projects that this market will amount to 10–45mtce per annum by 2000, with the great increase in nuclear-generated electricity which is also planned making the lower figure more realistic.

The iron and steel industry used 14.4mt in 1979, most of which was imported. The WOCOL Report sees this market reaching 17–20mtce per annum by 2000, although the problems of the French steel industry during the 1970s, and the challenge of Third World producers, make this figure seem optimistic. Other industrial plants consumed only 2mt in 1979, but under the 1980 energy programme coal is expected to substitute for oil in two-thirds of all industrial boilers, creating a market for 16mt by 1990. This compares with the WOCOL projection of 14–40mtce per annum by 2000. The same source forecasts residential and commercial use of coal, which stood at 5mtce per annum in 1979, reaching 4–10mtce per annum by 2000.

France is the world's second largest coal importer, with the USA (9.4mt), South Africa (7.8mt), West Germany (4.7mt), UK (2.6mt) and Australia (1.8mt) its main sources in 1981. The 1980 energy programme expects imports to reach 40mt by 1990, and the WOCOL Report projects a total of 38–115mtce per annum by the year 2000. Within this total, East European imports are expected to stagnate or decline. Transport facilities in France are adequate to handle increased coal imports, although substantial investments will have to be made to accommodate the higher projections. The main ports will be the deep-water terminals at Le Havre, Dunkirk and Fos, although some supplies to Lorraine will be routed through Rotterdam. Much inland transport of coal will be by barge, which should be facilitated by new high-capacity canals linking Fos to the Rhone. Construction of a high-capacity Rhine–Rhone link may also be necessary. The environmental problems of coal, although considerable, are unlikely to be highly controversial because of more intense opposition to nuclear power, except in areas with a high concentration of coal-using facilities.

France has been one of the more determined countries in making changes to its energy policy as a result of the 1973 oil crisis. Coal has been a major beneficiary of this, and consumption should rise rapidly, with production declining less quickly than once seemed likely. The WOCOL projections of coal consumption ranging between 48–125mtce per annum seem realistic, although low economic and energy demand growth are likely to make the upper level optimistic.

GREECE

In 1981, Greek production of lignite was 26.2mt, from surface mines at Ptolemais, Megapolis and Aliveri. Most was mined by the state utility, DEH, for immediate combusion in power stations, but some is used for fertiliser production and domestic heating. Output will increase substantially in the 1980s, tapping recoverable reserves of 511mtce and additional resources of 379mtce. Coal imports will also rise from the 0.5mt of 1979, mainly to supply new coal-fired power stations.

IRELAND

Ireland produced 0.1mt of coal in 1981, from mines in Connaught and Leinster.[41] Recoverable reserves amount to 55mtce, with additional resources of 40mtce, although exploitation is hampered by high ash contents. Construction of new coal-fired power stations will lead to a growth of coal imports from their present low level, perhaps to 2–3mt per annum by the 1990s.

ITALY

Coal production in Italy is on a minor scale, with 2.0mt of lignite extracted from Suchis, Sardinia, deposits in 1981.[42] Although production will expand to 3mt per annum, and other small deposits on the mainland will be tapped, recoverable reserves of only 15mtce severely restrict the possibilities. Italian coal consumption has remained at roughly the same level for fifty years, amounting to about 19mtce per annum in 1979. Imports of 18.4mt were mainly sourced from the USA (8.5mt), Poland (3.0mt), South Africa (3.8mt) and West Germany (2.9mt) and Australia (1.8mt). Most of this is used for metallurgical purposes, with some 2mtce per annum being burnt by the state utility, ENEL, for electricity generation. Although metallurgical use is expected to remain at a similar level for the rest of the century, ENEL consumption is forecast to reach 14mtce per annum by 1990. This will be achieved by conversion of existing oil-fired capacity and construction of 13 500MWe new capacity at coastal locations in Central and Southern Italy. By 2000, the WOCOL Report projects total coal consumption at 25.5–48mtce per annum, whilst the Shell company has forecast that imports will reach 38–40mt by 1990.

Although Italy's extreme dependence on imported energy, and public opposition to, and disarray in, its nuclear power programme will encourage a rapid growth in coal use, several obstacles remain. New port facilities will need to be rapidly constructed, although some use will be made of existing terminals for coking-coal imports, notably Taranto. Environmental difficulties may also occur, especially in already heavily-polluted industrial areas. Finally, lack of government co-ordination may slow the rate of expansion, as has been the case in other energy sectors.

LUXEMBURG

The Luxemburg steel industry imported approximately 2.7mt of coke and coking coal in 1979, mainly from West Germany. Demand is expected to remain at a similar level in the future.

NETHERLANDS

Although the Netherlands was once a large producer, the last mine closed in the 1970s.[43] Coal resources of 1.2 btce still exist in the Limburg field, but are too fractured and in too thin seams to be economically worked.

Coal imports amounted to 7.2mtce in 1981, of which 70 per cent was for iron- and steel-making, and 25 per cent for power generation. All new base-load stations are to be coal-fired, and coal is expected to provide 40 per cent of electricity needs by 2000, providing a market of 11.9–13.8mt per annum. Metallurgical use is unlikely to grow, but the Netherlands may be one of the first countries to introduce coal gasification as a means of 'diluting' natural gas from offshore fields or foreign sources to the calorific value of the nitrogen-rich gas of the Groningen fields, which are at present the main Dutch energy source. A Shell-Koppers demonstration gasifier is located at Moerdijk, and an Exxon catalytic pilot gasifier at Rotterdam. Use of *in situ* gasification techniques may also allow the Limburg deposits to be exploited.

According to the WOCOL Report, Netherlands coal consumption will reach 22.8–38.2mtce per annum by 2000 when, according to government plans, it will supply 20 per cent of energy needs. A policy of conserving natural gas reserves, and opposition to nuclear power, make these targets seem achievable, although the environmental problems of either coal combustion or gasification will be considerable in a densely-populated country. Most of the necessary imports will be shipped to Rotterdam, which handled 9.4mt of coal in 1979, although Amsterdam and Terneuzen are also coal ports. A new coal terminal at Maaslavt, Rotterdam, will have a capacity of 25mt per annum by the mid-1980s and will serve both Dutch and other European consumers.

NON-EEC COUNTRIES

Of these countries only Spain is a substantial producer, and this will remain the case. Their combined imports amounted to approximately 10mtce per annum in 1979, and the total is expected to rise substantially by 2000, especially in Spain and Sweden.

AUSTRIA

Austria produced 3.1mt of lignite in 1981, from the surface mines of West Styria, and imported 2.7mt of hard coal.[44] Recoverable reserves of

33mtce and additional resources of 36mtce give little scope for substantial increases in output.

After the rejection of nuclear power in a 1978 referendum, increased emphasis has been placed on coal as a source of electricity. At least 1 200MWe of coal-fired generating capacity will be installed in the 1980s, mainly using Polish coal from mines financed by Austrian loans, which may be transported by slurry pipeline. A power station may also be built in Eastern Austria to burn Hungarian lignite, although attempts to reduce dependence on COMECON sources of energy are likely.

FINLAND

Finland has no coal reserves, and imports totalled 5.7mt in 1981, mainly from Poland and the USSR. Some two-thirds was used for electricity generation and district heating, with the remainder consumed by metallurgical industries. Increased use of coal for electricity generation and commercial/residential purposes is expected, and the WOCOL Report has projected a consumption of 8.7–13.4mtce per annum by 2000.

NORWAY

In 1979, 260 000t of coal was mined in the Longyear coalfield, which is shared with the USSR, on the Arctic island of Spitsbergen.[45] The high-quality coking coal is mainly used in the ferro-metallurgical industries of North and Central Norway. Spitsbergen deposits at Svea may be exploited to offset a decline in Longyear, with the possibility of some coal being shipped to Oslo and South Norway, where approximately 0.9mt was imported in 1981. However, recoverable reserves of 18mtce and additional resources of 150mtce preclude large increases in production.

PORTUGAL

Portugal produced approximately 0.4mt of hard coal and lignite in 1981, and imported a further 0.3mt. Recoverable reserves of 22mtce and additional resources of 5mtce preclude any substantial rise in coal production, and coal imports are likely to increase by 2000.

SPAIN

Spanish coal production stood at 35.7mt in 1981, of which 20.9mt was lignite and 14.8mt hard coal.[46] This provided 16 per cent of primary energy needs. Hard-coal deposits – which often have very high sulphur contents – are concentrated in the Asturias Basin of North-West Spain, and lignite at La Corunna. Other deposits are also exploited at Andorva, Zaragoza and in Andalusia.

Poor geological conditions, labour problems and bad management have made Spanish coal some of the most costly in the world, with one state-owned producer, Hunosa, losing $270 million in 1979. Despite this, the country's recoverable reserves of 636mtce and additional resources of 3.3btce have led to a planned increase in output to 38mt per annum in 1987, which will mainly be surface-mined lignite and sub-bituminous coals. Most of the extra output will be used for electricity generation – 3 000MWe of new coal-fired capacity was announced in 1979 – although industrial use is also being encouraged by financial incentives. Imports of coking and steam coal, which stood at 8.3mt in 1981, are also expected to increase to almost 15mt per annum by 1986. A state-owned company, Carboex, was formed in 1980 to meet these requirements, either by direct purchase or joint production ventures overseas.

SWEDEN

Sweden presently produces a small amount of coal as a by-product of clay extraction, and reserves are negligible.[47] Coal imports amounted to 2.3mt in 1981, mainly for iron and steel production, although small quantities were used in cement and district heating plants.

A large proportion of future electricity-generating capacity will be coal-fired, especially after the present nuclear reactor programme is completed, whilst government incentives will encourage coal use in district heating and industrial plants. To meet the increased demand, improvements to the ports of Oxelosund, which presently handles most coking-coal imports, Gothenburg and Landskrona will be necessary, with the possible addition of a 300 000DWT capacity transhipment terminal by the 1990s, to serve all the Scandinavian countries.

The WOCOL Report projects a Swedish coal consumption of 17.1–25.9mtce per annum by 2000, although environmental factors, particularly the question of sulphur-dioxide emissions and acid rain (which is harming the ecology of Swedish lakes), and the effects of renewable energy and conservation programmes may prevent this total being reached. Official projections suggest that imports will reach 6–8mt per annum by 1990.

SWITZERLAND

Swiss imports amounted to 0.9mt in 1981 and an absence of indigenous energy sources may lead to increased use in the future. One study sees a potential for 7–9mtce per annum coal use by 2000, for electricity generation and synfuel production.[48] This would be imported by barge from Rotterdam or by barge/pipeline from Mediterranean ports.

8　The Third World

Until recently, coal production and use has been neglected in the non-Communist developing world. In 1977, their resources amounted to 203btce of hard coal and 27btce of lignite, only 2.3 per cent of the combined global total.[1] Recoverable reserves amounted to 48btce of hard coal and 17btce of lignite, or about 10.3 per cent of world reserves. In both cases, the low figures – in relation to geographical area – is almost certainly due more to lack of exploration than geological factors. Those reserves which are known to exist occur in about fifty developing countries and are exploited in about thirty. Third World production (including Yugoslavia) in 1979 amounted to 150mt of hard coal and 67mt of lignite – about 6 per cent of world output – of which more than half came from India, and most of the rest from Yugoslavia, South Korea and Turkey.

Although the energy crisis of the 1970s has focused attention on Third World coal, its development is hindered by several problems. One of these is lack of geological data and other information which is necessary to establish the viability of exploiting deposits. When this process has to begin from scratch, a period of 10–15 years is usually necessary before coal can be mined. Even when such information is available, large amounts of capital are required to finance mine construction and infrastructural development. In 1978, the development costs of a 15mt per annum project in Colombia were estimated at $850 million, while forecasts suggest that transport costs in less-developed countries will average 40–60 per cent of the delivered coal cost in the 1980s, compared to 10–30 per cent in developed countries. These high costs mean that coal projects will return only 10–15 per cent on capital invested, often considerably less than can be obtained from oil or other mineral projects.[2] And, even when capital is available, the necessary engineering and managerial expertise to ensure the successful development of coal may be lacking.

Despite these problems, the World Bank and the WOCOL Report suggest that output from developing countries will increase at a faster rate than the world average until the end of the century, when they

TABLE 8.1 Coal production prospects of developing countries

	Recoverable reserves[a] (million tce)	Production (million tce)				Annual growth rate (%)	
		1977	1980	1985	1990	1977–85	1985–90
Developed market economies	324 841	1 134.4	1 265	1 476	1 752	3.3	3.5
Centrally-planned economies	246 304	1 463.4	1 662	2 089	2 610	4.6	4.5
Developing countries							
Africa							
Botswana	3 500	0.2	0.2	0.3	5.0	5.2	75.5
Morocco	(96)	0.6	0.9	1.0	1.1	6.6	1.9
Mozambique	80	0.4	1.0	2.0	3.0	22.3	8.5
Nigeria	90	0.3	0.3	1.0	3.0	16.2	24.6
Zimbabwe	755	2.5	4.0	4.6	5.2	7.9	2.5
Swaziland	1 820	0.1	0.5	1.5	5.0	28.6	27.2
Zaire	(73)	0.1	0.1	0.1	0.1	0.0	0.0
Zambia	5	0.8	1.3	1.9	2.5	11.4	5.6
Tanzania	(360)	b	b	2.0	3.0	—	8.5
Burundi	NA	b	b	b	b	—	—
Algeria	(20)	b	b	b	b	—	—
Angola	(500)	—	—	—	2.0	—	—
Cameroon	(500)	—	—	—	—	—	—
Benin	NA	—	—	—	—	—	—
Egypt	(80)	—	—	—	—	—	—
Ethiopia	NA	—	—	—	—	—	—
Malagasy	(92)	—	—	1.0	2.0	—	—
Malawi	(14)	—	—	—	—	—	—
Niger	NA	—	—	—	—	—	—

TABLE 8.1 Coal production prospects of developing countries (Contd)

Sierra Leone	NA	—	—	—	—	—	—
Somalia	NA	—	—	—	—	—	—
Tunisia	NA	—	—	—	—	—	—
Total Africa	7 220	5.0	8.3	15.4	31.9	14.8	17.1
Asia							
Afghanistan	(85)	0.2	0.2	0.6	1.0	14.7	20.1
India	33 700	99.7	125.0	145.0	190.0	4.8	5.6
Indonesia	1 430	0.2	0.2	3.5	12.0	42.2	27.4
Iran	193	0.9	1.0	1.5	1.5	6.6	0.0
Pakistan	(1 375)	1.0	1.5	2.0	3.0	9.1	8.5
Philippines	(87)	0.3	0.3	1.6	4.0	23.3	20.1
South Korea	386	17.3	19.0	22.0	25.0	3.1	2.6
Taiwan	(680)	2.9	4.0	5.0	6.0	5.7	3.7
Turkey	793	7.4	9.8	13.0	16.2	7.3	4.5
Viet Nam	(3 000)	6.0	10.0	15.0	20.0	12.1	5.9
Thailand	(78)	0.2	0.5	2.0	6.0	33.4	24.6
Burma	(280)	b	b	b	b	—	—
Bangladesh	519	—	—	—	2.0	—	—
Brunei	(1)	—	—	—	—	—	—
Malaysia	(75)	—	—	—	2.0	—	—
Laos	NA	—	—	—	—	—	—
Total Asia	38 583	136.1	171.5	211.2	288.7	5.7	6.3

right The Third World 209

TABLE 8.1 Coal production prospects of developing countries (*Contd*)

Latin America							
Argentina	290	0.5	2.3	3.5	7.5	27.5	16.5
Brazil	8 098	3.5	6.4	10.0	20.0	13.9	14.9
Chile	162	1.2	2.0	2.5	7.5	9.6	24.6
Colombia	443	3.7	5.0	10.0	20.0	13.2	14.9
Mexico	875	6.0	6.7	8.0	9.3	3.7	3.1
Peru	105	b	0.2	0.3	0.4	*	5.9
Venezuela	978	0.1	1.2	5.0	8.8	*	12.0
Bolivia	NA	—	—	—	—	—	—
Haiti	(7)	—	—	—	0.3	—	—
Ecuador	(22)	—	—	—	—	—	—
Guatemala	NA	—	—	—	—	—	—
Honduras	(0.2)	—	—	—	—	—	—
Panama	NA	—	—	—	—	—	—
Total Latin America	10 952	15.0	23.8	39.3	73.8	12.6	13.4
Europe							
Yugoslavia	8 465	19.8	29.5	38.2	46.0	9.3	3.8
Total developing countries	65 219	175.9	233.1	304.1	440.4	7.1	7.6
Grand total	636 364	2 773.7	3 160.1	3 869.1	4 802.4	4.2	4.4

* Annual growth rate in excess of 50 per cent due to very low 1970 production base; NA – not available.
a Figures in parenthesis represent 'geological resources', – since no 'reserve' data available.
b Output below 0.1 million tce in 1977.

SOURCE World Bank, 'Coal Development Potential and Prospects in the Developing Countries', (1979).

should be responsible for approximately 10 per cent of world coal production[3] (see Table 8.1).

The strategy chosen for this development will vary between countries, depending upon the type and accessibility of deposits, the availability of markets and other factors. Most will be concerned to produce coal for local thermal or metallurgical use, to reduce dependence on imports or free domestically produced oil for export. A few, notably Botswana, Mozambique, Swaziland, Colombia and Indonesia, will produce coal for export. Others, such as South Korea, Taiwan and Hong Kong, will import coal as a cheaper and more reliable substitute for imported oil. Much of this expansion will be financed by newly-available loans from international financial institutions or capital from companies in the developed world, although political opposition to the latter will be strong.[4] The World Bank, in particular, intends to lend $100–200 million per annum for coal projects, which it expects to finance 17–35 per cent of all Third World investments in this field. Despite these inducements, Third World output will be dominated by present coal-producing countries until at least the 1990s. Four countries – India, Turkey, South Korea and Yugoslavia – should account for almost three-quarters of increased Third World output until 1985. By the end of the 1980s Brazil and Colombia can be expected to join them as major producers, whilst Botswana, Swaziland, Zimbabwe, Taiwan, Thailand, Indonesia, Argentina, Chile, Mexico, Venezuela and (possibly) Mozambique, Nigeria and the Philippines may reach production levels of 5mtce per annum by the early 1990s.

AFRICA

With the exception of South Africa, African production and use of coal is on a very limited scale and is unlikely to become substantial in the twentieth century. According to a WOCOL Report survey, production (excluding South Africa) should grow from 6mtce per annum in 1975 to 35mtce per annum in 2000, when consumption will be 24mtce per annum.[5] However, another survey by the French company, Charbonnages de France, suggests that although production will reach 50mtce per annum in 2000, it will be overtaken by consumption of 60mtce per annum.[6] The precise outcome will depend on international economic trends, the availability of aid for coal developments and political developments in Southern Africa, where economic and military tension constrained production in several countries during the 1970s. If this should continue, it will greatly impede any efforts to tap the large deposits of Botswana, Mozambique, Swaziland and Zimbabwe. The main African coal producers at present are:

Botswana

Although Botswana has recoverable reserves of 3.5btce of bituminous coal and additional resources of 10btce, production was only 0.4mt in 1981. This came from the Moropule mine, whose output is used for metal smelting and electricity generation. Other large deposits occur at Mmamabula and Mookane, but exploitation is hampered by the country's inaccessibility and their relatively low quality. Although the WOCOL survey forecasts that production will only be 0.7mtce by 2000, the World Bank projects an output of 5mtce per annum by 1990, and Charbonnages de France sees this rising to 10mtce per annum by 2000.[7]

Morocco

Morocco's production of anthracite from the Djerada mines reached 0.8mt in 1981, of which 10 per cent was exported, the rest being used for domestic purposes and electricity generation. Recoverable reserves of anthracite amount to 50mtce, and there are additional resources of 13mtce of lignite at Oued Nja. Mechanisation of the Djerada mines will increase output to 1mt per annum, and the WOCOL survey forecasts that coal production will be 2.2mtce per annum by 2000, with Charbonnages de France more optimistically projecting 5mtce per annum.

Mozambique

Production was 0.6mt in 1981, all from open-cast mines at Moatize and elsewhere in Tete province. With recoverable reserves of 240mtce and additional resources of 160mtce, production is expected to grow during the 1980s, although it will be constrained by infrastructural problems. Italian (Finsider) and East German (Transiter) steel interests are improving rail links from Moatize to a new coal terminal at Matola, and extensions are being made to Maputo coal terminal, to give a capacity of 5–6mt per annum for South African, Zimbabwe and Swaziland exports.

The WOCOL survey forecasts that output will reach 2.4mtce per annum by 1985 and 6.5mtce per annum by 2000, when exports will be 4.2mtce per annum. This is in line with World Bank projections of 3mtce per annum production by 1990, although less optimistic than the Charbonnages de France forecast of 10mtce per annum porduction by 2000, with half of this exported.

Niger

Production of approximately 0.2mt per annum of coal began in 1980, to supply the needs of the mining and metal industry.

Nigeria

Coal was once Nigeria's main commercial energy source, but output has been reduced by cheap oil supplies.[8] Recoverable reserves amount to 150mtce, with a further additional resources of 800mtce, but in 1982 production from the Enugu deep mines and Okaba (Benue) surface mine of the Nigerian Coal Corporation was only 0.2mt, far below capacity. Most of this was used in cement and other industries, and some 20 000t was exported to Ghana for railway use. Substantial investments were made in the late 1970s, but mechanisation caused operational difficulties, which were compounded by electricity shortages. Further investment is planned to allow production to reach 3.5mt per annum by 1985, although present difficulties and lack of markets make this target seem unrealistic. The World Bank sees production as reaching 3mtce per annum only by 1990, whilst both the WOCOL survey and Charbonnages de France see production being limited to 1.5mtce per annum or less by 2000.

Swaziland

The Mpaka mines of Swaziland Collieries (owned by the South African Anglo-American Corporation) produced 0.2mt of coal in 1982, of which part was exported via Mozambique. With recoverable reserves of 1.8btce of bituminous coal and additional resources of 3btce, future prospects are thought to be good. One likely development is exploitation of the Mhlume anthracite deposits, in which Shell have an interest. The World Bank forecasts that output will reach 5mtce per annum by 1990, whilst the WOCOL survey projects production of 6.1mtce per annum by 2000, of which 3.8mtce per annum would be exported.

Zaire

Zaire has recoverable reserves of 600mtce, but production amounted to about 0.1mt per annum in the 1970s, mainly for electricity generation and – to supplement approximately 0.3mt per annum of coking-coal imports – metal smelting. The World Bank sees no prospect of production increasing, although the WOCOL Report forecasts 2000 output of 1.4mtce per annum, with a similar level of imports.

Zambia

Zambia's output from its Maamba coalfield, near Lake Kariba, amounted to 1.3mt in 1981, used for power generation, industrial purposes and (blended with Zimbabwean coking coal) metal smelting.[9] Recoverable reserves are 24mtce, with 98mtce of additional resources. The World Bank forecasts that production will reach 2.5mtce per annum by 1990, with further rises projected by the WOCOL survey (3mtce per annum) and Charbonnages de France (5mtce per annum) by 2000.

Zimbabwe

Zimbabwe's coal production has been limited by political factors, and stood at 2.0mt of coal in 1981.[10] Recoverable reserves are 734mtce and additional resources total 5.8btce. Present output is entirely from underground and surface mines at the Wankie Colleries (also owned by Anglo-American) in the North-West of the country. Its good-quality coking coal is used by metallurgical and other industries, and considerable quantities are exported to neighbouring countries. Production should reach 7mt per annum by the mid-1980s, when the mines will be feeding an adjacent 1 280MWe power station. Extensive deposits which occur in other regions, particularly at Lubumbi and Sangwa, are also expected to be developed. Previous projections have been made obsolete

by the country's independence in 1980 and, if stable economic conditions persist, should be well above the 5.2mtce per annum forecast for
1990 by the World Bank.

Other countries with minor coal production at present are Algeria,
Burundi and Tanzania. The latter has recoverable reserves of 200mtce
(of which the Ruhuhu valley deposits are likely to be developed first) and
additional resources of 1.5btce. The World Bank projects that output
will reach 3mtce per annum by 1990, and the WOCOL survey forecasts
this level for 2000. Other countries which are expected to be producing
up to 2mtce per annum by 2000 are Angola and Malagasy, by which
time Algeria, Egypt and Tunisia may have become significant coal
importers.

ASIA

With the exception of India, one of the world's largest coal producers,
coal output and consumption in non-Communist Asian developing
countries is not large. In the absence of substantial domestic energy
sources, and given high rates of growth in many countries, particularly
in East and South-East Asia, coal consumption is expected to greatly
expand and to be met largely by imports. However, greater development
of domestic reserves by members of the Association of South East Asian
Nations (ASEAN) can be expected.[11] The WOCOL Report forecasts
that demand from East and South-East Asia will rise from under 25mtce
per annum in 1977 to 200–240mtce per annum in 2000, with especially
large increases in South Korea and Taiwan. However, the difficulties of
meeting a demand of this size – especially in the face of competition
from Japan – and the likelihood of rates of growth less rapid in the
past – in South Korea, GNP actually fell during 1980, compared to a
WOCOL assumption of 9.8 per cent per annum annual growth from
1975–85 – make these forecasts highly optimistic. Eventual import
levels may turn out to be less than half the figures quoted.

Afghanistan

Afghanistan has a small coal industry, producing approximately 0.2mt
per annum in the late 1970s.[12] Recoverable reserves amount to 66mtce,
with additional resources of 400mtce, and the World Bank forecast that

output would reach 1mtce per annum by 1990, although this has been put in doubt by the Soviet intervention of 1979 and subsequent fighting.

Bangladesh

Bangladesh has recoverable reserves of 242mtce, but no production at present. In 1979, imports amounted to approximately 1.8mt of coal, mainly from India. The World Bank projects that production, mainly from the Rajshahi deposits, may reach 2mtce per annum by 1990, and Charbonnages de France forecasts that steam-coal demand will reach 5mtce per annum by 2000.

Hong Kong

Hong Kong is a small importer of coal at present, but both the WOCOL Report and Charbonnages de France forecast that steam-coal imports for electricity generation will reach 5mtce per annum by 2000, mainly from China.

India

With a 1981 output of 130mt, India accounts for half of Third World coal production.[13] Recoverable reserves of 13btce and additional resources of 91btce give it the potential to become one of the world's major producers by the end of the century.

GEOGRAPHY

Some fifty-four coalfields occur in most parts of India but mining is concentrated in its Central and Eastern regions. Most seams are thick — 75 per cent are over 4.8m thickness — and shallow, with generally low sulphur but high ash contents. The main coal states are:

Bihar. This includes the Jharia, Karanpura and Bokaro fields, which produce 41 per cent of national output, including almost all of India's coking coal. In the Jharia field, past working practices, surface developments and underground fires have sterilised a large part of unexploited coking coal reserves.[14] Plans are in hand, with Polish help, to rationalise mines and remove some surface obstructions to allow these to be exploited.

West Bengal. Most of the large Raniganj field falls within this state, which produces 23 per cent of national output. This is mainly steam coal, which is used for electricity generation and industrial purposes in Calcutta and other centres.

Madhya Pradesh. This state contains most of the Singrauli field, and smaller fields at Hasda-Arand, Mand-Raigarh, Pench-Kanhan and Tawa Valley, which produce 20 per cent of national output. Major expansion of surface mining with Soviet aid is planned for Singrauli (both in Madhya and Uttar Pradesh), including the Jayant mine which will produce 10mt per annum by the late 1980s. New power-plant and coal-enrichment facilities will also be built.

Andhra Pradesh. Exploitation of bituminous steam coal in the Godavari Valley produces 9 per cent of national output. Development of a major industrial complex to make use of these deposits is planned for the 1980s.

Tamil Nadu. Most of India's lignite production of 3.2mt per annum is from the Neyveli field, where output is expected to reach 20mt by the late 1980s with East German aid. Other states with significant reserves include Assam, Gujarat, Maharashtra, Orissa and Uttar-Pradesh.

ORGANISATION

Overall responsibility for the Indian coal industry lies with the Coal Department of the Union Ministry of Energy. Research and development is carried out in the Central Mining Research Station and the Central Fuel Research Institute (CFRI). Coal production is dominated by the government-owned Coal India, formed by nationalisation in 1973. Its four subsidiaries – Bharat Coking Coal, Eastern Coalfields, Central Coalfields and Western Coalfields – account for 88 per cent of national output, and a further 9 per cent comes from Singareni Collieries of Andhra Pradesh, which it jointly owns with the state government. In 1979–80, Coal India made a loss of Rs910 million ($118 million). Remaining hard-coal output is from the captive mines of the Tata and Indian Steel companies.

Mining technology and organisation is backward in comparison with industrialised countries, although it has improved since nationalisation. Most mines are small and worked by manual or semi-mechanised room-

and-pillar methods. By the 1990s, a third of output is expected to be from fully-mechanised longwall faces – compared to 2 per cent at present – and a similar proportion from semi mechanised room-and-pillar faces. Surface mining, which produced 26 per cent of national output in 1979, is usually mechanised, although manual loading is still practised in many areas. This source is expected to provide 50 per cent of national output by 2000, when the present maximum depth of 60m should reach 200m. Both surface and underground operations were hit in the late 1970s by shortages of electricity and diesel fuel. Power cuts in 1979 led to estimated lost production of 6mt, and caused problems of flooding during the monsoon season due to lack of pumping capacity. Almost 20 per cent of West Bengal mines forced to temporarily close for this reason. Similar problems can be expected during the 1980s.

In 1978, India's mining workforce amounted to 660 000, which several observers see as excessive, although understandable in a poor Third World country. Plans for increased mechanisation – involving a doubling of productivity per manshift from 0.63 tonnes in 1980 to 1.25 tonnes in 1990 – will inevitably mean fewer workers and therefore conflict with labour unions. Industrial relations and absenteeism were major problems in the 1970s, with Coal India reporting that the latter factor caused lost production of 5mt in 1979, especially during the harvest season. Corruption and gangsterism also appears to be prevalent in the Eastern coalfields.[15]

MARKETS

Power generation, which consumed 28mt in 1978/79, should soon be the largest market for Indian coal, as electricity demand is slated to rise by 10 per cent per annum until the mid-1980s and by 6.5 per cent from then until 2000.[16] Although some of the increased capacity will be met by hydro and nuclear development, coal requirements are forecast at 54mt in 1983 and 215mt in 2000. In the past, problems have been experienced with commissioning new plant and co-ordinating the system, and these may produce a lower than projected rate of growth, especially if they coincide with economic troubles. Continued difficulties may also be encountered with the high ash content of Indian coals, which have already damaged several power-station boilers. However, transport needs should be eased by plans to build new capacity in the form of large, 2 000MWe, minemouth stations.

With a consumption of 35mt per annum in 1978/79, industry is presently the largest coal market. Projections suggest that demand will reach 94mt per annum by the year 2000, although this depends upon a high rate of economic growth being achieved. Supply shortages in the late 1970s, which led to the closure of some factories, may discourage reliance on coal as a fuel source. Metallurgical plants consumed 24mt of Indian coal in 1978/79 and forecasts of iron and steel production suggest that coal needs will reach 71mt per annum by 2000. In the 1980s, major expansion is planned in Bihar (Bokaro), Madhya Pradesh (Bhilai) and Andhra Pradesh (Vishakapatnam). New discoveries of domestic coking coal will help meet this demand, although increased use of directly-reduced iron processes may allow more steam coal to be used. India is one of the few countries with a largely coal-fired railway system and this market took 12mt in 1978/79, although it is expected to decline to 6mt by 2000. Indian coal exports have declined to 0.6mt in 1978/79, mainly to neighbouring countries. Continued domestic problems are likely to limit them to 1–2mt per annum until 2000. There will also be a continued need for coking-coal imports, which amounted to 1mt from Australia and Canada in 1979. These may be ensured by Indian investment in overseas deposits.

INFRASTRUCTURE AND ENVIRONMENT

The realisation of these optimistic demand forecasts will depend on improvements to transport facilities.[17] Indian Railways – which moves 78 per cent of coal production – has been overstretched for many years, and pithead stocks rose to 14mt in 1979. Difficulties have been compounded by the need to transport large amounts of coal waste, due to a lack of coal-preparation plants. New investments are being made, but some observers believe that a much more ambitious (and costly) programme is required in order to meet future needs. In 1979, for example, despite an estimated shortage of 30 000 wagons, only 13 000 were being made annually, of which a large proportion were merely replacements for old stock. As a result, coal has begun to be transported by road, at a great cost in money and scarce diesel fuel.

Although the environmental effects of production and use are as great as elsewhere, India's poverty has muted concern. Particular problems have been encountered with subsidence as a result of shallow room-and-pillar workings, and water pollution and land restoration in arid areas. Although plans to re-organise the Jharia coalfield, which involve the movement of whole villages, will have a great effect, environmental and social factors are unlikely to be a major constraint on future production.

COAL AND ENERGY POLICY

Coal is clearly destined to remain the major commercial fuel in the Indian economy for the foreseeable future, and there is no doubt that both production and use will greatly expand. Under the Sixth National Plan, output will reach 165mt per annum by 1985 and 400mt (286mtce) by 2000, and substantial investments are being made to reach these goals – amounting to Rs 17 billion ($2 billion) over the period 1980–5.[18] However, the problems of transport, labour relations and power supplies which caused output to stagnate in the late 1970s have not been entirely solved, whilst the country's heavy dependence on imported oil, and the economic difficulties which it causes, will remain for the foreseeable future. This will probably mean that targets will be downgraded during the 1980s. Nevertheless, by the 1990s, India will be a major force on the world coal scene, and both Coal India and domestic equipment manufacturers can be expected to play an important role as a sympathetic source of expertise and technology for other developing countries.

Indonesia

In 1979, Indonesian coal production was 0.3mt from deposits in South and West Sumatra, and Obilin, Borneo. Recoverable reserves of 234mtce and additional resources of 6.2btce will allow considerable increases in output. Expansion at Bukit Assam, in South Sumatra, is planned – with involvement by Shell – with the coal transported to new power stations at Suralaya and elsewhere in West Java.[19] Some 12GWe of new coal-fired generating capacity is planned for the end of the century, and the World Bank forecasts that output will reach 12mtce per annum by 1990.

Iran

The Iranian revolution led to a fall in coal output from the approximately 1mt per annum level of 1977. Recoverable reserves amount to 193mtce per annum and the World Bank projects that production will reach 1.5mtce per annum by 1990, although this will depend upon political factors.

Israel

Israel imported approximately 1mt of coal in 1979, and plans to increase this to 10mt per annum by 1990, largely for electricity generation.[20] Two 1 400MWe coal-fired plants are under construction, and existing units – and industrial facilities – will be converted from oil to coal.

Malaysia

Malaysia has coal resources of 260mtce, and the World Bank forecasts that output will reach 2mtce per annum by 1990. The WOCOL Report projects a steam-coal demand of 3mtce per annum, mainly for electricity generation, by 2000, compared to a Carbonnages de France forecast of 5mtce per annum.

Pakistan

Coal production in Pakistan was 1.1mt in 1981, from deposits at Lakhra, Sind and Baluchistan. Recoverable reserves stand at 394mtce, and the World Bank forecasts that output will reach 3mtce per annum by 2000.

Philippines

Philippines coal output stood at 1mt in 1979, mainly of lignite. Recoverable reserves amount to 64mtce, and further deposits exist on Luzon and Mindanao. Ambitious production targets of 4.7mt per annum by 1985 have been set, compared to a World Bank forecast of 4mtce per annum output by 1990. Construction of new coal-fired power stations and conversion of cement, sugar and other factories is also expected to lead to increased coal imports. Carbonnages de France projects a steam-coal consumption of 10mtce per annum by 2000, compared to a 16mtce per annum forecast for the same year by the WOCOL Report.

Singapore

The WOCOL Report, and Charbonnages de France, forecast that Singapore will import 5mtce per annum of steam coal by 2000.

South Korea

After a rapid expansion following the Korean War, South Korean coal production declined during the 1960s and 1970s, amounting to 21mt in 1981, with a further 10mt of imports.[21] Recoverable reserves, mainly of anthracite, stand at 116mtce, with additional resources of 1btce.

Present output is entirely deep-mined anthracite, exploited in poor tectonic conditions mainly from the provinces of Cholla Namdo and Kangwanda. The state-owned Dai Han Coal Corporation accounts for a third of production. Substantial investments are being made in new facilities and the World Bank forecasts that output will reach 25mtce per annum by 2000. Over half of coal sales are for residential purposes, with most of the rest for industrial uses. Only 10 per cent of output is presently used for electricity generation, but this market is expected to grow rapidly. Charbonnages de France forecasts that total steam-coal demand will reach 40mtce per annum by 2000, whilst the WOCOL Report forecasts 74–92mtce per annum for the same date, when coal will account for 30–50 per cent of installed generating capacity. Most of this will be imported, partly from mines with direct South Korean investment in Australia and Canada. However, the unrest and severe recession which hit South Korea in 1979–80 cast some doubt on the performance of its economy in the 1980s and beyond.

Taiwan

Taiwan produced approximately 3mt of sub-bituminous coal in 1979, when recoverable reserves stood at 109mtce. The World Bank forecasts that output will reach 6mtce per annum by1990, although the WOCOL Report believes that it will have declined below present levels by 2000. Forecasts of steam-coal demand for 2000 are wildly disparate, with Charbonnages de France quoting a figure of 15mtce per annum and the WOCOL Report 55–66mtce per annum.

Thailand

In 1981, Thai lignite output was 1.5mt, from Mae Mah and other deposits in Northern Thailand. Production is planned to reach 4mt per annum by 1984, tapping recoverable reserves of 34mtce, and the World Bank forecasts that other resources may be tapped to raise output to 6mtce per annum by 1990. The state-owned utility, EGAT, intends to expand lignite-fired generating capacity from 150MWe in 1979 to 1 425MWe in 1990. By 2000, steam-coal demand is projected to reach 10mtce per annum by Charbonnages de France, although the WOCOL Report believes that it will only be 1mtce per annum.

Turkey

Turkish coal production amounted to 4mt of hard coal and 16.5mt of lignite in 1981.[22] The state-owned coal company, Turkish Coal Enterprises (TKI), plans to substantially increase output, taking advantage of recoverable reserves of 757mtce (mainly lignite) and additional resources of 10.2btce (mainly bituminous).

Hard-coal mining is concentrated in the Zonguldak Basin of North-Central Turkey, with output sold mainly to local industrial and metallurgical users. Unless continuing exploration is successful, coking and steam-coal imports will increase as production fails to keep pace with demand, which TKI estimates will reach 16.7mt per annum by 1987. The main lignite deposits are in Western and Central Anatolia, most of which is burnt in power stations, providing 10 per cent of Turkish electricity. Major expansion is planned, particularly from a 20mt per annum surface mine at Elbistan, being developed with West German help. TKI has forecast a market of 120mt per annum by 1987, but this – as with hard-coal forecasts – is likely be undermined by economic problems and political instability. By contrast, the World Bank predicts that total output will double to reach 16.2mtce per annum by 1990.

LATIN AMERICA

Although coal production has been extensive in the past, Latin America is today only a minor element in the world coal scene. According to a market survey commissioned for the WOCOL Report, coal production should reach 35mtce per annum by 1980, 97.7mtce per annum by 1990 and 146.9mtce per annum by 2000, with actual tonnages far higher because of the dirty nature of much Latin American coal. Imports are forecast to rise from 11.2mtce per annum in 1980 to 60.9mtce per annum in 2000, and exports from 7.7mtce per annum to 24.4mtce per annum.

Latin America is one of the fastest growing and most economically advanced areas of the Third World, which – together with the oil-created trade deficits of many countries – should encourage greater production and use of coal. However, many individual countries can be expected to have substantial economic problems during the 1980s which, together with a traditional political hostility to the development of natural resources by American business, may make the necessary investments difficult. One factor which may alleviate this, and other problems, is the further development of Latin American economic integration, such as the Latin American Energy Organisation (OLADE).[23]

Argentina

Coal production in 1979 was 1.4mt of bituminous coal from the Rio Turbio mines of South Argentina, operated by the state-owned YCF.[24] Recoverable reserves amount to 117mtce and there are additional resources of 3.4btce. Much of this is lignite, both at Rio Turbio and in Santa Cruz. Output was hit in the late 1970s by capital shortages, falling demand and limited capacity at the port of Rio Gallego, where the coal is shipped to consumers, which are mainly power stations. As a result, imports of approximately 1mt per annum were necessary, mainly from the USA and Poland. Production is projected to reach 4.8mtce in 1985 and 8.6mtce per annum by 2000, by which time imports will be 11.2mtce per annum, mainly of metallurgical coal. In view of the industry's present difficulties, these forecasts seem optimistic.

Brazil

Brazil is a significant coal user, but its 1981 production of 5.3mt failed to meet local demand.[25] With recoverable reserves of 910mtce and additional resources of 11.4btce, Brazil plans to expand the industry considerably during the 1980s and beyond.

Over 70 per cent of recoverable reserves are in the East Parana Basin of Rio Grande and Santa Caterina, mainly of sub-bituminous quality. However, large deposits of good-quality coking coal were discovered in Rio Grande during 1979. Other coal seams are found in Rio Fresco and Goias and, in the form of lignite, in the Upper Amazon Basin. This region is largely unexplored, and further large deposits are expected to be found.

Mining is carried out by private and state-owned companies – of which the largest are CRM and COPEIMI – but the government has indicated its intention of denationalising the industry and opening it up to foreign capital. Polish and French interests are already active in the country. In 1978, 60 per cent of coal consumption was for electricity generation and 25 per cent for metallurgical uses. Cement plants were another important market, and coal is expected to be the only energy source in this industry by the mid-1980s.

Under the 1979 coal expansion plan, which includes financial incentives for both mining and use, indigenous coal production is expected to reach 35mt per annum by 1985. To achieve this target an investment of $7.3 billion in forty-two new mines is envisaged. Coal consumption is projected to grow to 29mtce per annum in 1990 and 45.3mtce per annum in 2000, when about 40 per cent will be domestically provided. The bulk of the additional imports will come from the USA, which will supply at least 45mt in the course of the 1980s. Much of this extra consumption will be in the burgeoning steel industry, although coal will face competition from charcoal (derived from fast-growing eucalyptus trees) and direct-reduction processes (although some of these will use coal gas) for the market. Coal-fired electrical capacity should also rise from 1 000MWe in 1979 to 25 000MWe by the late 1980s. Coal gasification for fertiliser and chemical production is also expected to expand, and a Koppers-Totzek gasifier will be operating in Rio Grande by the mid-1980s.

The realisation of Brazil's ambitious targets for the coal industry will depend very much on the country's economic performance, which will be hampered in the 1980s by a huge foreign debt and other problems. Production plans may also be handicapped by infrastructural problems, lack of skilled labour, inadequate resource assessments and the low quality of many coal seams. Nevertheless, it is already clear that Brazil will be a major producer and consumer of coal for the rest of the century.

Chile

Coal production in Chile amounted to 0.8mt in 1981 and the country exported small amounts of coal to neighbouring countries. Recoverable reserves are estimated at 920mtce and additional resources are put at 3.5btce. The main economic deposits are of high-ash sub-bituminous coal in the Magellanes Basin, in Southern Chile (which extends into the Rio Turbio field of Argentina), where future development, will be concentrated. Deposits are also found near Valdivia, and in the Arauco Basin, where production is expected to end in the near future. Production is projected to reach 3.1mtce by 1985 and 5.8mtce by 2000, with continuing exports on a small scale.

Colombia

With recoverable reserves of 1btce and additional resources of 7.7btce, Colombia has the potential to become a major coal producer and exporter, and to build the largest coal industry in Latin America.[26] In 1981, however, production amounted to only 5.5mt, and considerable difficulties must be overcome before major expansion is possible.

Colombia's coal deposits were created in the troughs and depressions associated with the uplifting of the Andean mountains and are scattered across much of the country. The main reserves now being worked, and with the potential for large increase in production, are in the Eastern Cordillera from Bogota to the Venezuelan border; in the Western Cordillera around the city of Medellin; along the Caribbean coast and in the Valle area of South-West Colombia. Generally, the coal is in thin seams and is not amenable to surface mining. Particular problems to its exploitation are caused by lack of infrastructure, with rail, waterway and port facilities being inadequate. The most feasible solution lies in a considerable upgrading of the waterways network and the development of Caribbean ports to handle exports from nearby sources, with some additional capacity to ship coal from inland deposits.

General responsibility for coal matters in Colombia lies with the Ministry of Mines and Energy, whilst coal development is mainly under the control of Carbones de Colombia SA (Carbocol), owned by the state oil and mining companies, and other government bodies. Until recently, production has been hampered by a multitude of small producers, many unlicensed, whose technology has included the use of child labour. This has been altered by the $3 billion Intercor (an Exxon subsidiary)-Carbocol development of the El Cerrejon coking-coal deposit on the Guajira peninsula of North-East Colombia, whose production should reach 15mt per annum by 1990.[27] The deposit will be surface-mined and requires construction of a 150km railway and a new port at Bahia Portete. Exploration agreements have also been signed with Brazilian, Romanian and Spanish state bodies, a consortia of West German steel companies and numerous US companies. As with El Cerrejon, these are mostly concerned with coking-coal deposits.

About 29 per cent of current output is sold to metallurgical users and a further 15 per cent is used for electricity generation. The rest is consumed by small industries and domestic buyers. Exports amounted to approximately 0.2mt in 1979, mainly to other Latin American countries for coking purposes. The WOCOL Report projects coal production at 18mtce per annum by 1990 and 32mtce per annum by 2000, of which 35–40 per cent will be exported. This compares with a World Bank forecast of 20mtce per annum output by 1990. Infrastructural problems may make this target difficult to achieve, but further development of coastal deposits for export to the USA and Europe, or advances in Latin American economic co-ordination, may provide the investment needed to overcome them.

Mexico

Coal was once Mexico's main energy source, but in the twentieth century it has been overshadowed by oil, and production in 1981 was only 7.6mt, which provided 5.3 per cent of national primary energy needs.[28] Plans are in hand to increase production, making use of the country's recoverable reserves of 1.5btce (mainly bituminous) and a further 1.7btce of additional resources. Most deposits are found in Coahuila, especially in the Muzquiz and Monclova areas; Barrancas; Oaxaca and Sonora. The Coahuila seams are of low–medium-quality bituminous coal, whose coking quality is reduced by a high ash content, making 0.6mt of imports necessary for blending purposes in 1979.[29] Metallurgical industries are the main coal users, although electricity consumption will be boosted by completion of a 1 200MWe coal-fired station at Rio Escondido.

The WOCOL Report projects that coal production will rise to 55mtce per annum in 2000, when a further 19.2mtce per annum of inports will be necessary. Coal should then provide 14.3 per cent of primary energy needs. However, these figures may be rendered optimistic by the impact of cheap oil and gas, including the use of the latter for direct reduction of iron ore (a process pioneered in Mexico). Their exaggeration is underlined by the World Bank forecast of 9.3mtce per annum output by 1990, and a Carbonnages de France figure of 40mtce production in 2000.

Peru

Peru has a small quantity of recoverable reserves at Alto Chicama, and additional resources estimated at 8.68mtce. In 1979 there was some small production for local markets, but plans to develop output at Alto Chicama for electricity and metallurgical uses has been hampered by problems of water supply. This places in doubt previous forecasts of 3.4mtce per annum production by 1985 and 6.5mtce per annum by 2000.

Venezuela

Venezuela has recoverable resources of 134mtce, mainly in the Eastern Cordillera and the Zulia Basin of North-West Venezuela, with a further 8btce of additional resources.[30] Limited production comes from the mines of Carbones de Zulia, which produce a low-ash bituminous coal, suitable for coke production when blended. Production is projected by the WOCOL Report to reach 7mtce per annum by 1985 and 16mtce per annum by 2000, when some 1–3mtce per annum of imports will be necessary. This compares with government projections of 4–5mt per annum output by 1985, on the basis of a $1 billion investment programme. However, investments in coal may seem less attractive in future than the development of the huge heavy-oil deposits of the Orinoco Basin.

9 The Future of Coal

THE PAST OF COAL

Coal was the foundation of the modern world. It powered the engines of the Industrial Revolution, and warmed, lit and transported the populations of the sprawling cities which they created. Its chemical and physical properties made it a basic raw material for the steel, chemical and other industries. The mines from which it was won made vast fortunes for their owners, and employed millions who – with their families – changed the political and social development of their nations.

Coal also had its price. Countless thousands of miners died, were maimed or were simply worn out. Suffering, or the fear of it, was accompanied in times of depression by poverty, shared by whole communities – collective experiences which often marked them off from neighbouring villages or areas. The landscape was scarred by waste tips and other intrusions, while in the cities grime and smoke blanketed buildings, bringing death by bronchitis and respiratory conditions to many of their inhabitants.

The costs of coal ensured that when alternative energy sources became available, non-economic factors reinforced their financial attraction. Thus, in the twentieth century, coal lost all its traditional markets except for metallurgical coke and was only saved from virtual extinction by the rapid growth of electricity generation. Production fell precipitously in many countries and stagnated in others, with the exception of the Communist bloc whose steady output growth meant that the world total did not suffer an absolute fall. In the West, a two-tier coal industry was created, with a relatively prosperous coking sector, and a steam sector producing a low-value commodity subject to fierce competition from oil, natural gas and nuclear power, and therefore operating on low margins.

The problems of coal were in marked contrast to the rapid expansion of other energy industries in the years between 1950 and 1973. In particular, oil, with its giant Middle Eastern reservoirs, seemed to offer virtually limitless supplies of cheap energy, without undue environmental impacts. And, for the future, there was the promise of low-cost electricity from nuclear power stations. The warnings of coal producers that the rapid expansion of energy demand would one day outstrip supply, and that the coal industry should therefore be supported as an insurance policy against this occurrence, went unheeded.

COAL AND THE ENERGY CRISIS

In retrospect, the oil bubble began to subside in 1970 when US output peaked, and the price rises of the early 1970s reflected this fact. However, it was the Yom Kippur war of 1973, resulting in an Arab embargo on oil supplies to some Western countries, and OPEC's quadrupling of prices, which brought the message home – although in the illusory form of fear of absolute scarcity rather than adjustment to changing economic conditions. Immediately, coal usage increased as, despite a rise in its own price, coal became suddenly competitive with fuel oil. And, as it became clear that high energy prices were to be a permanent feature of the world economy, many countries and organisations formulated policies to increase coal production and use in the medium–long term.

This dramatic change in the fortunes of the world coal industry during the mid-1970s is almost unique in economic history. Coal is not the only growth industry of one era to become the laggard of the next, only to be revitalised again by changing market conditions – but it is by far the largest. Within a few years, it was expected to overcome the disadvantages of its recent and historical past – the image of a depressed industry, with deleterious effects on staff morale and the quality of recruitment; under-capitalisation and obsolete technology; the tensions between management and labour which these often caused; an inadequate reserve base due to cutbacks in exploration; public opposition to its environmental and social impact, based on past experience; and its dependence on volatile and unreliable markets. As one study commented, 'a traditionally backward industry must be suddenly transformed into a modern, technologically advanced one'.[1]

In view of these obstacles, it is not surprising that coal expansion in the 1970s did not occur on as rapid a scale as many governments and other interested parties had hoped. However, any disappointment was tempered by setbacks in other energy industries. Oil and gas prices were clearly going to remain high and supplies subject to disruption for political reasons. Public opposition to nuclear power slowed construction programmes and in some countries produced an effective moratorium on new building. Renewable energy sources were some years from full commercialisation and many policy-makers were sceptical as to whether they would ever be a major energy source. By the end of the decade, many observers saw coal as the only energy source whose output could grow sufficiently quickly to meet anticipated deficiencies in energy supply, would reduce the dependence of many countries on energy imports, and could be processed into suitable forms for the different sections of the energy market. At the same time, the lesson of the 1970s had been learnt, and it was equally clear that, if these goals were to be achieved, unprecedented political and economic support had to be provided to the coal industry and its consumers. Notable decisions to provide this support were made by Western leaders at their 1980 summit conference, the Communist states in their COMECON discussions and national plans, and the World Bank with regard to its lending for coal projects in Third World nations.

THE WOCOL REPORT

The most comprehensive, and optimistic, forecasts of major coal expansion have been made in the Report of the World Coal Study Group, which included representatives from coal producers, coal users and governments from OECD and other countries.[2] For each member country, the Report considered two scenarios – a high coal, in which world energy demand grew at 2.5 per cent per annum, and a low coal, in which it grew at 1.7 per cent per annum. When aggregated, these figures suggested a more than doubling of world coal demand to 2000 in the low coal case and a trebling in the high-coal case. The Report argued in considerable detail that this rapid increase in demand could be supplied. World coal reserves are enormous, and the technologies of production and transport generally well-proven and readily available. Manufacturing capacity is, or could easily become, sufficient to provide the necessary equipment, for which capital would be readily available. The environmental effects of coal could be kept at an acceptable level by existing control measures, and should eventually be reduced by the introduction of new coal technologies, which will also increase the thermal efficiency of coal-energy systems. The main obstacles to a major expansion of coal production are organisational, which could be overcome by determined government action to reduce uncertainty about the future and to encourage co-ordinated investment in the complex' 'coal chains' which are needed to bring supplies from mine to market.

The message of the Study – which was, naturally, not displeasing to its members – was that coal was available as a 'bridge to the future', which could provide most of the world's incremental energy needs to the twenty-first century without requiring any fundamental change in existing energy, economic or political systems. In urgent tones, the Report concluded that:

> A massive effort to expand facilities for the production, transport and use of coal is urgently required to provide for even moderate economic growth in the world between now and the year 2000. Without such increases in coal the outlook is bleak.[3]

The publication of the WOCOL Report in 1980 had an immediate, and considerable, impact on energy policy-makers and has undoubtedly served its purpose of encouraging further coal development. The message has also been reinforced by other analyses, which stress the potential importance of coal in the world economy. According to one expert:

> Coal development will bring various favourable effects, including the improvement of international balance of payment situations, through a controlled build-up of world trade, reorganisation of large-scale infrastructures, repletion of production and transportation facilities, the development of related technology and the creation of direct and indirect effective demand through the development of coal-related industries. Thus this offers a real opportunity to break the economic stagnation that exists in both developing and industralized countries.[4]

The WOCOL Report has been extensively criticised on methodological grounds – particularly the fact that the national forecasts on which it is based make different assumptions as to the future course of world oil prices, and the resulting income and substitution effects – and because it forecasts an end which is clearly desired by its members and proceeds to 'prove' that this can be achieved. Nevertheless, in the early 1980s its projections are likely to be justified, as coal production and use rises rapidly as a result of high oil prices. During this period, unused supply and use capacity can be brought quickly into operation, and coal substituted for other fuels in the areas where this is relatively easy. However, in the medium–long term, many of the assumptions and arguments contained within the WOCOL Report are open to serious question.

ECONOMIC GROWTH AND ENERGY DEMAND

Energy forecasting is a particularly difficult task, especially in the unique world of the 1980s, where the highest projections of future growth in energy are now less than the conventional assumptions of the 1970s. As one study has commented:

> Energy markets are too complex, energy products are too diverse, and the unknowns are too many to be reduced confidently to one description of future events. The growth of the economy, policies designed to affect demand, the resolution of important environmental debates, and the evolution of energy prices will have dramatic effects on the level and composition of future energy demands.[5]

The WOCOL Report forecasts are based on what it terms moderate rates of economic and energy demand growth. Thus, estimates of GNP growth to 2000 for WOCOL-participating states range from 2.2 per cent to 4.6 per cent per annum, and it is assumed that this will be accompanied by a 25 per cent decline in the amount of energy used per unit of economic activity. However, since compilation of the Report in the late 1970s, the world has endured a major recession, and opinions as to future levels of economic growth and energy conservation have changed significantly in the West, the Communist bloc and the Third World. Growth forecasts have been persistently marked downward as a result of the effects of continued petroleum price rises and OPEC limitations on output; the anti-inflationary measures taken in many countries; structural problems in both national and international economies and political instability, as with the Iran–Iraq war which began in 1980. Many observers believe that these factors will remain important throughout the 1980s and, possibly, the 1990s. As one major study pessimistically concluded:

> If present trends continue, the world in 2000 will be more crowded, more polluted, less stable ecologically, and more vulnerable to disruption than the world we live in now. Serious stresses involving population, resources and environment are clearly visible ahead.[6]

Although major coal expansion would help to mitigate some of these problems, their deep-rooted nature will influence future coal prospects more profoundly than coal's contributions to solving these problems. High energy prices, and the economic burden of energy imports for many countries, also seem likely to produce greater improvements in energy conservation than the WOCOL Report suggests. This is especially true of the USA, where decontrol of oil and gas prices has proceeded more rapidly than seemed possible in the 1970s.

Within the limited growth in energy demand which will occur to the year 2000, the WOCOL Report assumes that coal will successfully compete with other energy sources – or will replace them because of shortfalls in supply – to take a very much larger market share. Thus, coal will increasingly take over the role of oil – whose output is expected to peak before 2000, while the volume available for export stagnates or declines – in industrial and utility bulk-heat markets and, by the end of the century, liquid fuels and chemical feedstocks. However, although a considerable amount of substitution will clearly occur, its scale may be limited by the effects of high oil prices in commercialising previously uneconomic oil supplies. These include both conventional oil reservoirs and new supplies from heavy oil, shale oil or tar sand deposits, as well as enhanced recovery from existing wells. Similar considerations apply to gas supplies, which may limit the movement of coal into industrial markets, and its use as a source of SNG.

Other factors may also determine the extent to which coal can substitute for oil. One of the most marked features of the world economy in the 1970s and early 1980s was a pattern of recession induced by sharp and substantial oil price rises, leading to oil 'gluts' and more stable pricing policies. The letter encouraged economic growth, but this was choked again as oil producers took advantage of increased demand to raise prices. Coal and other energy sectors faced difficulties over investment and the opening up of new markets during recessions – when financing was difficult, energy demand reduced and oil was freely available – and were unable to react quickly enough to the short-lived periods of expansion. The result was that, for all the talk of developing coal and other energy sources in the 1970s, the world was as dependent upon oil for its energy supplies in 1981 as it was in 1971. Similar problems may continue to stymie the coal industry during the next decade.[7]

Within this broad picture, the future use of coal will be greatly influenced by developments in its main market, electricity generation. In the past, electricity demand has grown more rapidly than overall energy demand, and the WOCOL Report assumes that this will continue to be the case. Yet, as electricity is an inefficient means of using energy, it may be reserved in future for premium uses, as is already advocated by some observers. Some evidence for such a trend was apparent in the late 1970s and early 1980s, when electricity demand stagnated and utilities drastically reduced their future building programmes. The consequent over-supply of generating capacity has also limited the scope for construction of new coal-fired plant in many countries.

This slowdown in electricity growth will intensify competition between coal and nuclear power for new base-load capacity. Although most utilities believe that coal is more costly, this disadvantage has been offset by public opposition to, and the technical problems of, the nuclear industry, particularly since the Harrisburg accident of 1979. However, increased awareness of the risks of coal production and use may change this situation in the future. Ironically, some of the new coal technologies which may reduce these risks could also reduce coal demand from the electricity sector by using it more efficiently.

COAL PRICES

In practice, the scale of coal expansion to the year 2000 will be primarily determined by its price, which – because of the higher handling and use costs of coal – must offer a significant differential over competing fuels.

On the supply side, large reserves of coal exist which can be mined for some years at a constant, or only slightly increasing, real cost. In the medium–long term, however, the more accessible, and easily mined, deposits will be exhausted, and higher-cost reserves will be exploited (see Table 1.6). This will involve a change in the balance between surface and underground mines, the working of deeper seams in both, and a transition from room-and-pillar to longwall techniques in the latter. These changes will partially counterbalance the effects of increased automation in reducing the high labour content of coal-mining costs, so that the industry will remain sensitive to wage and salary settlements or improved working conditions. Many countries will also impose stringent environmental and safety controls on coal production, transport and use, with similar effects on costs and productivity levels to those which occurred in the USA during the 1970s. Finally, the infrastructural costs of opening up new coal deposits will also increase, although in some cases these will be partially compensated for by a decline in ocean-transport costs as larger ships are brought into service.

In the case of coal demand and its effect on prices, one critic of the WOCOL Report has noted that it assumes 'a classical model of competition where capacity at each stage of the supply chain adjusts smoothly to changes in demand and price. . . . This model has no room for market imperfections, due to monopoly power or physical limitations to capacity expansion, or for the resulting economic rents (the difference between the price that a factor of production can command and its cost of supply)'.[8]

In fact, many opportunities to extract economic rents are, or could be, available in coal-energy systems. Coal consumers purchasing their requirements from individual mines on long-term contracts may be unable to resist price increases, particularly where their facilities have been adapted to particular types of coal. Transport enterprises, especially railways and ports, may also monopolise coal movements from and to individual areas, and their freedom to charge 'what the market will bear', rather than cost-related prices, is already a controversial issue in the USA. Coal marketing and broking organisations, particularly the multinational oil companies, may also be in a position to increase prices. Finally, governments may also seek to derive higher revenues from the coal industry, either by increasing taxation levels or encouraging a rise in export prices.

The extent of these opportunities has been emphasised by a 1980 study of inflation and energy prices:

> Higher coal prices in the last seven years largely reflect rising costs of production. The fact that prices have not risen as strongly as oil prices suggests that little economic rent is being captured. But the extent to which this could alter is suggested by natural gas price changes in the last twelve months, which are beginning to bring internationally traded prices up to the thermal equivalent of oil. If strong energy markets re-emerge with a renewal of economic growth during the 1980s, there might be pressure on coal producers to capture economic rent up to the point where coal just retains competitiveness with oil in the markets (power generation and heating oils) where it can supplant oil; to reach this point coal prices would have to rise faster than oil prices. And if countries attach sufficient importance to the development of large domestic coal reserves, or coal imports which are more secure than oil imports, the competitive limit on coal price increases would be higher.

Whereas such a demand analysis points to a possible increase in coal prices, some supply analyses suggest that coal price increases will be restrained by an abundance of low-cost coal and the attraction of new suppliers into the market if excess profits are made. This may be true if there is a wide range of countries supplying internationally traded coal, but if the international market becomes dominated by a few suppliers or in any way comparable to OPEC's control of oil, the pressure to extract maximum value to the domestic economies by charging high coal prices may be hard to resist. The point is that there is considerable room for coal prices to rise relative to oil before coal's competitive edge is lost.[9]

Some evidence for this view was provided in 1981, when a combination of increasing demand and disruption of supplies from Poland, the USSR and the USA produced the first coal sellers market since 1975. This produced substantial increases in the price of Australian steam coal – which had already doubled in price during 1979–80 – as well as supplies from other countries.

COAL RISKS

The view of the WOCOL Report that the health and environmental hazards of major coal expansion can be kept within acceptable limits may also be optimistic. Existing data indicates that, in normal circumstances, coal is the most hazardous of energy sources, with underground coal mining being one of the most dangerous of all occupations, and coal use having deleterious effects on public health. Its ecological and environmental impact is also of major proportions and is becoming international in nature. Evidence that atmospheric transport of coal pollutants contributes to acid rain in countries far from the point of use has already created international tension, and may eventually represent a significant constraint on the levels of coal use in many areas. Similar considerations may apply in the case of carbon dioxide, although firm evidence of deleterious effects on the earth's climate will take many years to amass.

To date, public opposition to coal expansion on health and environmental grounds has been minimised by the greater degree of controversy surrounding nuclear power, and – in areas other than those directly affected – by a familiarity with the impact of coal, making it seem more a 'known' quality. However, as production and use rises in the 1980s and beyond, more attention will be focused on this area, and performance of the coal industry scrutinised more carefully. Expansion in the 1990s and beyond may be severely constrained if the promises made by coal producers and consumers in the late 1970s and early 1980s have not been fully adhered to. At present, the coal industry also benefits from an assumption that the extent of its impact, although onerous, is relatively well defined. However, any major coal expansion may produce qualitative as well as quantitative changes in this impact, again leading to a reappraisal of support for this policy. One example of this may be the possible tendency of coal-using facilities, such as power stations or coal-conversion plants, to cluster around production units in so-called 'coalplexes'.

NEW TECHNOLOGIES

There is a widespread assumption, in the WOCOL Report and elsewhere, that the development of new coal technologies will significantly improve the competitiveness and reduce the risks of coal in the future. This applies particularly in the case of fluidised-bed combustion and gasification technologies for bulk heat production, these and MHD for electricity generation, and coal-conversion plants for liquid fuels and chemicals. Such views are primarily based on the fact that the new processes generally have a higher thermal efficiency than those presently in use, and reduce some of the costs currently associated with environmental controls in the USA and other countries.

Although there is considerable evidence to support these views, it is also true that many comparable industries – particularly nuclear power – have found the large-scale introduction of new technology as complex and costly as that envisaged for coal to be a more difficult, and slower, process than first imagined. It is already clear that major engineering and materials problems have still to be overcome in the initial development of many new coal technologies, and that further obstacles, including shortages of raw materials, such as bulk oxygen supplies or skilled manpower, may hinder full commercialisation. Solutions to these will almost certainly inflate the already high cost estimates for new coal plant, possibly making them less attractive investment choices than the new technologies of other energy industries. Finally, some of the new coal technologies may create as yet unidentified, or underestimated, hazards to health and the environment.

THE POLITICS OF COAL

In the 1960s and early 1970s, the relative decline of the coal industry was reflected in its shrinking political importance. For many countries, coal policy was primarily a question of handling the social and economic problems created by its demise. And, even where coal retained some economic significance, it was seen primarily as an adjunct of the electricity or steel industries.

Since 1973, the growing emphasis on 'energy policy', and the renaissance of coal production, have taken the coal industry to the centre of the political stage. The move has brought many advantages, particularly that of government support for its continuing expansion. This has been reflected in many areas, including the provision of capital, statutory or voluntary persuasion of utilities and other consumers to increase coal use and the encouragement of coal imports and exports. Some attempts have also been made to introduce greater co-ordination between administrative bodies, particularly in the USA. Such actions reflect the views of the WOCOL Report, which concludes that governments, 'by providing the confidence and stability required for investment decisions, by eliminating delays in licensing and planning permissions, by establishing clear and stable environmental standards, and by facilitating the growth of free and competitive international trade' will be of crucial importance in determining the future scale of coal expansion.[10]

Coal's growing political importance has also had its disadvantages. Its conspicuousness has encouraged greater scrutiny of its record and plans, and intensified both local and national opposition to its expansion, primarily on environmental and social grounds. To date, this opposition has not been of crucial importance because of more intense controversy surrounding the subject of nuclear power, but the situation may change in the 1980s as coal expansion continues and cumulative impacts occur. In some countries, notably the UK, the renaissance of coal has a direct party political importance, so that the Conservatives are lukewarm to further expansion.

The politics of coal will be of especial importance in four countries – the USA, Australia, South Africa and Canada – which will dominate future world coal trade. All have Federal political systems, which allow considerable scope for blocking actions by local groups or decentralised government units, despite pro-coal policies at a national level. This factor has been particularly important in limiting coal expansion in the USA, and may become increasingly significant elsewhere. These countries have also experienced enduring debates on the merits of exporting raw materials, rather than capturing the higher value associated with processed products. Associated with this is a feeling that foreign investment, and long-term contracts, reduce their independence and force citizens to bear environmental and other costs for the benefits of foreigners. A further argument, particularly important for strategic reasons in South Africa, is that present output levels should be limited in order to maintain reserves which will increase in value over the years. Should these attitudes become more prevalent, the presently favourable policies of the Federal governments in these countries may change. This would follow a pattern familiar for other raw materials, in which initial government encouragement to a young industry is followed by greater regulation and control when it has matured and its impacts are apparent.

In geographically large coal-producing countries, future expansion may also become entangled with questions of regional development and equity. In the USSR, coal expansion means massive investment in Siberia, whereas priority to other energy sources, particularly nuclear power, would favour its European regions. Conversely, in the USA, coal expansion will be concentrated in Western states, further enhancing their growth prospects at the expense of other parts of the country.

COAL INTERNATIONAL

In recent decades, coal has been a national or regional energy source. Only coking coal has been internationally traded on a significant scale, and most steam coal has been consumed within a few hundred miles of its production.

Most forecasts suggest that this situation will change by 2000. Thus, the WOCOL Report projects a 300–500 per cent increase in world coal trade during this period, with most of the increase attributable to steam coal. Although the bulk of coal output will continue to be consumed in national markets, a large number of countries will be dependent on coal imports for a significant share of their energy requirements.

This internationalisation of coal will, in many respects, make it a more flexible and dependable energy source, with consumers less dependent on a small number of suppliers and having available a greater variety of coals for purchase. In geopolitical terms, it should also reduce the dependence of OECD, and some COMECON, countries on OPEC oil imports and help reduce international tension over raw material supplies.

However, the trend will also create new problems. Short-term flexibility may be purchased at the long-term cost of being locked into particular supply chains, vulnerable to disruption or the exercise of monopoly power by 'Western' organisations or governments, as anxious to gain the maximum revenues from their resources as any Arab sheikhdom from its oil. As the disruption of the world coal trade caused by the Polish industrial unrest of the early 1980s demonstrated, exports from COMECON or Western countries may also be equally liable to disruption by political or social unrest. They may also be influenced by diplomatic factors, as with the reluctance of some countries to import South African coal. In the long run, the coal market may also follow the oil market in being dominated by the co-ordinated policies of a few large coal producers and exporters and/or the multinational oil companies. This economic internationalisation of coal will also be accompanied by a similar trend in its environmental impact, so that coal production and use may be increasingly influenced by foreign-policy considerations. This is becoming the case in Northern Europe and the USA, where acid rain created by sulphur-dioxide emissions from coal combustion seems to have caused serious damage to the environment in other countries and has created considerable international tension. As coal use grows, the scale of this problem can be expected to increase. Similar tensions will also occur if the link between carbon dioxide and a greenhouse effect is proven, especially if – as is likely – political solutions were slow in coming.

COAL: BRIDGE TO WHERE?

The WOCOL Report, and other forecasts before it, have suggested that world coal output should be almost tripled to act as a 'bridge to the future'. In fact, many factors – a slowdown in world economic and energy demand growth; higher than expected availability of other energy sources; infrastructural and environmental problems of coal development; uncertainty surrounding coal-related investments and public opposition – suggest that the WOCOL projections are over-optimistic. A doubling of world coal output by 2000 is probably the upper limit of what is possible, although further expansion may occur beyond this date.

Even if the WOCOL targets were reached by 2000 – or the decade thereafter – it is open to question whether this would be desirable. What would happen is a substitution or supplementation of one fossil fuel – oil – by another, coal. This could merely bequeath the problems of one generation to the next, so that the world in the year 2003 or 2023 or 2043 faces a 'coal crisis' comparable to the 'oil crisis' of 1973. That is, it will have become dependent on a relatively low-cost energy source to the neglect of other energy technologies, and is suddenly confronted with rapid increases in price, and painful economic and social adjustment, as a result of a transition to higher cost reserves and/or the full extraction of economic rent by private organisations and governments alike. In addition, the cumulative environmental impacts of coal may become so serious that expansion must be halted and choices of alternative energy developments constrained.

It is likely that the world energy system would be as unbalanced by reliance on coal and oil as it will be for oil alone. Nor it is likely that nuclear power, despite the considerable increase in output which can be expected in the 1980s, will fully overcome the technical, economic and political difficulties which hampered it in the 1970s. Given this situation, there is an urgent need for the world to develop more diversified energy systems, using a variety of energy sources – with an increasing component of renewable energy – in as flexible and efficient a manner as possible. The balancing act which needs to be performed in coming decades is the maintenance – or slight increase – of conventional energy supplies to shore up existing centralised, inflexible and relatively inefficient energy systems, without starving new energy resources of much-needed investment. The WOCOL Report is right that coal expansion is the key to this process – but wrong as to the level of increase which is necessary, or desirable. Expansion on the scale advocated by WOCOL would tip the balance too much towards existing systems, and too much against emerging technologies. Rather, a moderate expansion of coal output – with a doubling as an upper, and temporary, limit – is necessary for the next 20–50 years, after which time coal, like oil, can be increasingly reserved for the premium uses to which its high energy content makes it best suited. In short, coal is a bridge to the future, but a limited, utilitarian structure, like the Ironbridge over the English Severn, or 'the rude bridge that arched the flood' at the battle of Concord – not a colossal, endless construction, like those which span the Firth of Forth or San Francisco Bay. There are bridges and bridges.

Postscript

During the 1980s the optimism created in the coal industry by the WOCOL Report and the oil price rises of 1979–80 was tempered by a realisation that world energy futures were more complex, and the problems of massive coal expansion more formidable, than had been assumed.

The economic recession which began in 1979 proved more enduring than first anticipated and, although a tentative recovery began in 1983, few observers believed that the serious nature of world economic problems would allow a return to the high growth rates of the pre-1973 era for many years, if ever. The combined effects of recession, more efficient energy use and the development of non-OPEC oil production also led to an oil 'glut' in the early 1980s, weakening prices and creating a widespread belief that supply would remain ahead of demand for much of the 1980s and even (a small minority believed) beyond. These developments undermined the basic assumptions of WOCOL and other highly optimistic coal projections, which were predicated on high world economic growth rates and oil prices.

The effects of this new situation were apparent in the early 1980s when, despite a considerable increase in world coal trade between 1979–82, world coal supply was higher than demand and prices obtained by coal exporters fell markedly. By 1983 South African coal was selling at $36 per tonne in Europe, a price at which very few exporters were making a profit. Most observers expected surplus coal capacity to persist – being exacerbated in the 1980s by the development of Third World exporters such as Colombia – with a consequently depressive effect on prices and, ultimately, new investment. Although low prices stimulate coal use, the effects of recession and development of other energy sources in many countries gave little hope that this surplus will be wiped out by rapidly growing demand until at least the 1990s.

Coal expansion was also overshadowed by environmental factors.[1] Concern at the effects of acid rain continued to grow in North America and Europe, and more stringent sulphur and nitrous oxide emission standards were introduced in several countries. Those introduced in 1983 by West Germany – which estimated that 10 per cent of its forests were being destroyed by acid rain – had an estimated cost to the electricity industry of up to DM6 billion, and put 1–2p/kWh on to electricity prices. In Britain the NCB's plans to develop the North-East Leicestershire (Belvoir) coalfield were stymied on environmental grounds, although one pit eventually went ahead. These and other measures are likely to increase the cost of producing and using coal in the 1980s.

One country which was greatly affected by these changes was the USA.[2] Although output rose to 823mt in 1982 both domestic consumption and exports fell – in the latter case to little over 80mt compared to 100mt in 1981. Although construction of several new export terminals at Baltimore, Newport News and elsewhere eased the serious shipping problems of the early 1980s high rail transport rates continued to plague all producers. The refusal of the Reagan administration to intervene or finance investment in new infrastructure did much to counteract other pro-coal measures – such as deregulation of oil and gas prices, the roll-back of environmental legislation, and release of federally owned coal bearing lands. Its cutback of support for coal conversion projects also dashed hopes that gasification or liquefaction would provide significant coal markets before the end of the century. One of the few which survived – the Great Plains gasification plant – is anticipating losses of up to $1 billion, and possibly more, in its first decade of operation, which augurs badly for future developments.[3] Many coal buyers are also anticipating increases in real coal prices during the course of the 1980s. Hence, while output will expand considerably in coming years, there is little evidence that it will rise to the levels predicted by more optimistic forecasts.

In the USSR a new output target of 770–800mt per annum has been set for 1985, reinforced by large wage increases for miners, increased investment and some imports of equipment.[4] Almost all the incremental output of the 1980s will come from surface mines, whose output is planned to expand from 269mt in 1980 to 400mt in 1990, mainly from the Kansk-Achinsk, Ekibastuz and South Yakutia coalfields. However there have been continuing problems with labour shortages, poor equipment and geological difficulties in the Donets and Kuznetsk fields which have made it impossible to reach the 1983 output target of 723mt. Coal use is being expanded by conversion of oil- or gas-fired power station and other boilers although efforts are also being made to increase the efficiency of coal (and all energy) use, with limited success. The combined effects of supply shortfalls and limited conservation will limit the future level of Soviet coal exports.

In China too earlier hopes continue to be downgraded as the scale of its economic problems, and the effects of continued political uncertainty, become apparent.[5] Output recovered from its 1981 fall to 625mt in 1982 but was well short of the latest 700mt target for 1985. To achieve this the sixth five-year plan announced in 1983 included 28 large new mines, some of which would use foreign capital. The first of these was investment by Occidental in a 15mt per annum Shanxi development, half of whose output will be exported.

By 1983 the Polish coal industry had recovered much of the ground it lost in the early 1980s, although neither output nor exports seemed likely to regain their 1979 level for a number of years.[6] Official targets were revised to 214mt production by 1990 and 255mt by 2000, although the large amounts of coal wasted in the development of the Lublin field and other mines, and the continuing possibility of labour and social unrest, will make them difficult to achieve.

A shadow fell over Australian coal in the early 1980s, and in 1982 several New South Wales mines were closed.[7] Exports fell slightly in 1982 and in 1983 Japanese steel companies negotiated a $15 per tonne reduction in price for their Australian coking coal imports. A combination of strikes, concentration amongst coal producers – notably in the takeover of Utah International by Broken Hill Proprietry – and increased government intervention all led Japanese importers to reappraise their dependence on Australia and to begin a diversification to other sources of supply. The election of a Labour government in 1983, with its proposal for a 'resource rent tax' on mining companies also alarmed the domestic coal industry. Nevertheless the quantity and quality of Australian coal resources – supplemented in 1982 by the discovery of large coking coal deposits in South Australia's Arckaringa basin – and the continuing improvement in coal export infrastructure suggested that these problems might slow, but would not halt, the industry's expansion.

Australian producers were undercut in European and other markets by South African coal, which provided their lowest-cost supplies.[8] Although exports dipped to 27mt in 1982 an increase to 80mt by 1995 was forecast, encouraged by the liberalisation of export quotas in the early 1980s. The Mineral Bureau also sees a tripling of total output to 250mt in the year 2000, although this assumes high economic growth and continued political stability.

Like South Africa Canada benefited from the Japanese desire to diversify from Australian and US supplies.[9] Investment in new capacity – as with the $1 billion Quintette, British Columbia development which will export 6.3mt annually to Japan by the mid 1980s – railways and ports and good industrial relations make it likely that exports will more than double in coming years. Domestic consumption is also rising strongly.

West European coal use grew less quickly than anticipated (although imports still increased substantially) and the British, West German and French mining industries were all heavily subsidised and over-producing by 1983.

In the UK the NCB faced an ever-deepening crisis.[10] Its 1982–83 production was much higher than real demand, which was in-flated by CEGB purchases for stock and export sales at low prices – dumping, in the eyes of some observers. Demand in 1983–84 is expected to fall even further and stocks increase to well above the 52mt held by the NCB and coal users in early 1983. Prospects for significant growth in electricity or industrial demand during the remainder of the 1980s are poor and a likely increase in the latter market during the 1990s may be compensated by a fall in CEGB purchases. The forecasts prepared for the Sizewell nuclear reactor inquiry by the Department of Energy saw a coal market of 100–140mt per annum by 2000, with the lower end of the range appearing to be more likely – a far cry from the optimistic figures of 'Plan For Coal'.

Stagnant demand and high stocks are compounded by the large number of high-cost pits which the NCB operates. Thirty of these accounted for 90 per cent of its 1981 deficit and three lost over £10 million annually. Already low margins are also likely to be squeezed by CEGB pressure for a new coal pricing structure, which was partially reflected in the renegotiated understanding between the two organi-sations in late 1982. According to CEGB evidence at the Sizewell inquiry these pressures meant that the NCB deficit might reach £1 billion per annum, compared to the almost £500 million of 1982–83. This prospect was unacceptable to the Conservative government which, after the embarrassing volte-face of 1981, imposed a new target of break-even by the late 1980s and appointed Ian McGregor as NCB Chairman. (In the early 1980s, as Chairman of British Steel, he had greatly reduced steel capacity.) His new appointment, and the general NCB strategy of reducing high-cost capacity and its labour force in favour of low-cost capital intensive production such as that from the new Selby field – expected to be five times more productive than the national average – was fiercely opposed by the NUM, now led by left-wing President Arthur Scargill, and a succession of industrial disputes in 1982–83 suggested that the 1980s would be a period of conflict between management and unions.

In West Germany too coal consumption fell in 1982, despite an increase in coal use for electricity generation.[11] Rising stocks and mounting government subsidies produced demands for a more liberal import policy, while the stringent sulphur emissions standards introduced in 1983, and the cutbacks in the ambitious coal conversion programme of the early 1980s greatly reduced optimism about the future. The socialist government, elected in France in 1981, gave great encouragement to the French coal industry, with an energy programme which envisaged domestic output rising to 30mt per annum by 1990, providing half of anticipated coal use.[12] By 1983, however, rising losses at CdF – which was losing Fr161 on every tonne produced – and lower than anticipated increases in energy demand led to a reversal of the plans.

Japan also downgraded its forecasts of coal imports as its economic growth slowed, and the future of energy-intensive industries became more problematic. Its estimated 1990 requirements fell to 130mt. Importers took advantage of world coal over-supply to renegotiate lower prices with Australian and other producers, and to diversify their sources of supply.

The evidence from almost all these countries is that rapid (as opposed to moderate) increases in coal production and consumption is unlikely to occur in the 1980s, and perhaps to the end of the century, unless world growth rates return to the levels of the 1960s – which most economists consider to be unlikely. Thus a BP forecast concluded that non-Communist world coal consumption would double between 1980 and 2000 only if developed countries enjoyed 3 per cent per annum growth, and developing countries 5 per cent, for the remainder of the century.[13] Several analyses – notably that by the CEGB for the Sizewell nuclear reactor inquiry – have also argued that coal prices will rise over coming decades to reflect the increasing marginal costs of moving from surface to underground production, which would also dampen increases in coal demand.[14] Other studies, however, have concluded that rapid coal expansion is still possible, although in many cases these arguments reflected views of what the future ought to be like – high economic growth and a rapidly expanding world coal trade to maintain a downward pressure on oil prices – rather than logical developments of early 1980s conditions.[15] These were that the world economy seemed likely to grow at a slow rate for the 1980s at least, that coal was only one amongst a number of energy options in most countries, that coal supply remained vulnerable to infrastructural and other problems and that coal expansion was often opposed (whether justifiably or not) by environmentalists. While there is not doubt that world coal production and trade will increase markedly in the 1980s and 1990s little has happened to change the conclusion of the first edition that this growth will proceed at a slower rate than envisaged in more optimistic forecasts.

Notes and References

For general accounts of coal see:

B. Grosling, *World Coal Resources* (London: *Financial Times*, 1980).

H. M. Lee, *Published Plans and Projections for Coal Production, Trade and Consumption* (London: IEA Coal Research, Economic Assessment Service, 1977).

Organisation for Economic Co-operation and Development, *Steam Coal: Prospects to 2000* (Paris: Organisation for Economic Co-operation and Development, 1978).

C. Simeons, *Coal: Its Role in Tomorrow's Technology* (London: Pergamon, 1978).

World Coal Study, *Coal: Bridge to the Future* (Cambridge, Ma: Ballinger, 1980).

World Coal Study, *Future Coal Prospects* (Cambridge, Ma: Ballinger, 1980).

Also copies of *Coal Age*; *Coal Mining and Processing*; *Coal Week*; *Colliery Guardian*; *Glückauf*; *International Coal*; *World Coal* and *World Coal Letter*.

CHAPTER 1   COAL: ORIGINS AND EXPLOITATION

1. A. Raistrick and C. E. Marshall, *The Origin and Nature of Coal* (London: English Universities Press, 1952); J. Gibson, 'Coal: an introduction to its formation and properties', in G. J. Pitt and G. R. Millward, *Coal and Modern Coal Processing* (London: Academic Press, 1979).

2. A. K. Matveev, 'USSR Geologists publish new map of world coal resources', *World Coal*, (1977) pp. 55–60.

3. D. Murchison and T. S. Westoll, *Coal and coal-bearing strata* (Edinburgh: Oliver and Boyd, 1968).

4. G. O. Argyll (ed.), '*Proceedings of the Second International Coal Exploration Symposium*' (San Francisco: Miller-Freeman Publications, 1979); M. Mitchell, 'Modelling geological uncertainty', *World Coal*, 6 (1980) pp. 55–7.

5. R. M. Davidson, *The Molecular Structure of Coal* (London: IEA Coal Research, Technical Information Service, 1980); D. W. Van Krevelen, *Coal: typology-chemistry-physics-constitution* (Amsterdam: Elsevier, 1961); J. W. Larsen (ed.), *Organic Chemistry of Coal* (Washington D. C.: American Chemical Society, 1978).

6. G. C. Smith and A. C. Cook, 'Coalification paths of exinite, vitrinite and inertite', *Fuel*, 59 (1980) pp. 641–6.

7. S. Jasienko, 'The nature of coking coals', *Fuel*, 57 (1978) pp. 131–46.

8. I. Evans and C. D. Pomeroy, *Strength, Fracture and Workability of Coal* (London: Pergamon, 1966).
9. C. Karr, *Analytic Methods for Coal and Coal Products* (London: Academic Press, 1978).
10. 'Reading the Composition of Coal', *EPRI Journal*, 5 (1980) pp. 6–11.
11. World Energy Conference, *Survey of Energy Resources 1980* (London: World Energy Conference, 1980).
12. M. J. Modelevsky *et al.*, 'Coal reserve classifications used in different countries', and J. J. Schanz, 'The classification of coal resources', in M. Grenon (ed.), *Future Energy Supply for the World Energy Balance*, (London: Pergamon, 1978).
13. ICF, Inc., *Coal Resource Information* (Palo Alto, Ca: Electric Power Research Institute, 1980).
14. World Energy Conference, op. cit.
15. S. M. Cassidy (ed.), *Elements of Practical Coal Mining* (New York: American Society of Mining, Metallurgical and Petroleum Engineers, 1973); National Research Council, *Coal Mining* (Washington, D.C.: National Academy of Sciences, 1978); Office of Technology Assessment, *The Direct Use of Coal* (Washington, D.C.: US Government Printing Office, 1979) Ch. III.
16. Hittman Associates, *Underground Coal Mining: An Assessment of the Technology* (Palo Alto, Ca: Electric Power Research Institute, 1976); I. C. F. Statham, *Coal Mining Practice* (London: Caxton, 1978).
17. C. Simeons, *Coal: Its Role in Tomorrow's Technology* (London: Pergamon, 1978).
18. 'New Developments in Surface Mining', *Coal Age*, 85 (1980) various articles pp. 72–121.
19. P. Wood, *Less-conventional underground mining* (London: IEA Coal Research, Technical Information Service, 1980); National Coal Board, *Mining Beyond 2000* (London: National Coal Board, 1977).
20. D. Jackson, 'Hydromining comes of age at Kaiser', *Coal Age*, 85 (1980) pp. 54–63; A. E. Gontov, 'Hydraulic mining in the Soviet Union', *World Coal*, 3 (1977) pp. 55–60.
21. S. J. Scheiderman, 'Thick and steep seam mining', *World Coal*, 6 (1980) pp. 30–3.
22. S. J. Curl, *Underground transport in coal mines* (London: IEA Coal Research, Technical Information Service, 1978).
23. J. R. Chadwick, 'MINOS', *World Coal*, 6 (1980) pp. 36–9.
24. S. J. Curl, *Underground coal gasification* (London: IEA Coal Research, Technical Information Service, 1979); G. J. Lamb, *Underground Coal Gasification* (Park Ridge, NJ: Noyes Data Corporation, 1977).
25. 'Coal preparation', *World Coal*, 5 (1979) various articles pp. 21–37; 'Fine coal cleaning', *Mining Engineering*, 32 (1980) pp. 1213–40; W. P. Bethe, 'Impact of changing raw coal properties and market trends on coal preparation', *Glückauf*, 116 (1980) pp. 445–9.
26. H. M. Lee, *The Future Economics of Coal Transport* (London: IEA Coal Research, Economic Assessment Service, 1980).
27. P. James, 'Twice as much coal by 2000', *Railway Gazette International*, 137 (1981) pp. 127–9.

28. Office of Technology Assessment, *A Technology Assessment of Coal Slurry Pipelines*, (Washington, D.C.: US Government Printing Office, 1978); R. A. Hill and J. D. Pitts, 'Slurry pipelines in the 1980's', *Combustion* (1980) pp. 41–6.
29. H. M. Lee, *The Long Run Economics of the Ocean Transport of Coal* (London: IEA Coal Research, Economic Assessment Service, 1978); P. James, 'Coal holds key to shipping prosperity', *Transport*, 1 (1980) pp. 67–8; H. P. Drewry, Ltd, *The Growth of Steam Coal Trade* (London: H. P. Drewry (Shipping Consultants), 1981).
30. World Coal Study, *Coal: Bridge to the Future* (Cambridge, Ma: Ballinger, 1980).
31. R. Zajackiewicz, *The Labour Force in Mining* (Geneva: International Labour Office, 1980); The President's Commission on Coal, *The American Coal Miner* (Washington, D.C.: US Government Printing Office, 1980).
32. J. Gavrenta, *Power and Powerlessness* (Oxford: Oxford University Press, 1980).
33. The President's Commission on Coal, op.cit. p. 7.
34. P. J. Palumbo, 'The future coal labour trends', *American Mining Congress Journal*, 66 (1980) pp. 31–44.

CHAPTER 2   COAL USE: TECHNOLOGIES AND MARKETS

1. For an overview see, M. A. Elliott (ed.), *Chemistry of Coal Utilisation* (New York: Wiley, 1980); I. Berkovitch, *Coal–energy and chemical storehouse* (Redhill, UK: Portcullis Press, 1979); G. J. Pitt and G. R. Millward (ed.) *Coal and Modern Coal Processing* (London: Academic Press, 1979); Office of Technology Assessment, *The Direct Use of Coal* (Washington, D.C.: US Government Printing Office, 1979) Ch. III.
2. A. L. Dainton, 'The combustion of coal', in G. J. Pitt and G. R. Millward, op. cit.; A. Culp, *Principles of Energy Conversion* (New York: McGraw-Hill, 1979).
3. G. F. Morrison, *Combustion of low grade coal* (London: IEA Coal Research, Technical Information Service, 1978).
4. Babcock and Wilcox, Inc, *Steam, its Generation and Use* (New York: Babcock and Wilcox, 1980); Central Electricity Generating Board, *Modern Power Station Practice* (London: Pergamon, 1971); S. Eriksson *et al.*, *Economic and Technical Criteria for Coal Utilisation Plant, Part IV: Power Generation* (London: IEA Coal Research, Economic Advisory Service, 1977).
5. N. Jenkins, 'Combined heat and power', *Energy Policy*, 8 (1980) pp. 169–73; Department of Energy, *Combined Heat and Electric Power Generation in the UK* (London: Department of Energy, 1979).
6. Comptroller General of the USA, *The Fluidised-Bed Combustion Process* (Washington, D.C.: US General Accounting Office, 1980); 'Fluidized-bed boiler', *EPRI Journal*, 4 (1979) pp. 6–13.
7. M. Simon-Tov *et al.*, *Technology assessment for an atmospheric fluidised-bed combustion demonstration plant* (Oak Ridge, Tn: Oak Ridge National Laboratory, 1979).

8. Comptroller General of the USA, *Magnetohydrodynamics* (Washington, D.C: US General Accounting Office, 1980); A. Sullivan, 'MHD coal power goes on the grid', *Coal Age*, 8 (1979) pp. 147–51.
9. P. James, 'Continuing interest, but patchy commitment to m.h.d.', *Electrical Review*, 206 (1980) pp. 29–30.
10. L. M. Pruce, 'COMs: boiler fuel of the future?' *Power*, 124 (1980) pp. 108–10.
11. 'Raising steam-coal power examined', *Motor Ship*, 60 (1980) pp. 47–9.
12. American Institute of Chemical Engineers, *Coal Processing Technology* (New York: American Institute of Chemical Engineers, 1979).
13. J. Gibson, 'Carbonisation and coking', in G. J. Pitt and G. R. Millward (ed.), op. cit.; Metals Society, *Coal, coke and the blast furnace* (London: Metals Society, 1978); S. Jasienko, 'The nature of coking coals', *Fuel*, 57 (1978) pp. 131–46.
14. N. Avram *et al.*, 'World trends in the reduction of coke consumption in iron making', *Metallurgica*, 28 (1976) pp. 213–21; K. G. Beck and J. Meckel, 'Extension of coking coal resources by new processes of coal preparation and carbonisation', *Glückauf*, 115 (1979) pp. 456–8.
15. Metals Society, *The Steel Industry in the 1980s* (London: Metals Society, 1980); R. Hodson, 'A new iron process lights the steelmaker's interest', *Financial Times*, 24 July 1980.
16. E. J. Hoffman, '*Coal Conversion*' (New York: Wiley, 1979); L. A. Anderson and D. A. Tillman, *Synthetic Fuels from Coal* (New York: Wiley, 1979); M. L. Gorbaty *et al.*, 'Coal science: basic research opportunities' *Science*, 206 (1979) pp. 1029–34.
17. D. Fishlock, 'West German faith in nuclear heat', *Financial Times*, 21 November 1980.
18. A. L. Dainton, 'Gasification of coal', in G. J. Pitt and G. R. Millward, op.cit.; S. Eriksson *et al.*, *Economic and Technical Criteria for Coal Utilization Plant, Part II: Gasification Processes* (London: IEA Coal Research, Economic Advisory Service, 1977).
19. W. L. Lom and A. F. Williams, *Substitute Natural Gas* (London: Applied Science Publishers, 1976).
20. 'Coal gasification for electric utilities', *EPRI Journal*, 4 (1979) pp. 6–13; M. E. Lackey, *Summary of the R & D effort on open-cycle coal-fired gas turbines* (Oak Ridge, Tn: Oak Ridge National Laboratory, 1979).
21. J. D. Homgren *et al.*, 'Dispersed coal-powered fuel cells for electric utilities', *International Power Generation*, 3 (1979) pp. 26–9.
22. National Coal Board, *Liquid Fuels from Coal* (London: National Coal Board, 1978); J. Owen, 'Liquefaction of coal', in G. J. Pitt and G. R. Millward, op.cit.; S. Eriksson *et al.*, *Economic and Technical Criteria for Coal Utilisation Plant, Part III: Liquefaction Processes* (London: IEA Coal Research, Economic Advisory Service, 1977); J. E. Schiller, 'Composition of coal liquefaction products', *Hydrocarbon processing*, 56 (1977) pp. 147–52.
23. D. Fishlock, 'Germany's plans to enrich the world's coal', *Financial Times*, 22 October 1980.
24. J. Owen, 'The coal tar industry and products from coal', in G. J. Pitt and G. R. Millward, op.cit.

25. World Coal Study, *Coal: Bridge to the Future* (Cambridge, Ma: Ballinger, 1980) Ch. 3.
26. D. F. Hemming *et al.*, *The Economics of Coal-Based Electricity Generation* (London: IEA Coal Research, Economic Advisory Service, 1979); W. D. Marsh, *Economics of Electric Utility Power Generation* (Oxford: Oxford University Press, 1981); 'Nuclear generation costs compared', *Nature*, 286 (1980) pp. 753 and subsequent correspondence.
27. M. Prior, 'Myths and mysteries of electricity from coal and nuclear power', *Coal and Energy Quarterly*, 23 (1979) pp. 24–34.
28. Congressional Research Service, *Federal Government Incentives to Coal and Nuclear Energy* (Washington, D.C.: US Government Printing Office, 1979).
29. M. Prior, *The Supply of Energy to Industry* (London: IEA Coal Research, Economic Assessment Service, 1979).
30. Exxon Corporation, *World Energy Outlook* (New York:Exxon Corporation, 1979).
31. World Coal Study, op.cit.
32. D. F. Hemming *et al.*, op.cit.
33. M. Prior, *The Supply of Energy to Industry*, op.cit.; D. Olliver, *Oil and Gas from Coal* (London: Financial Times Business Information, 1981).
34. M. Prior, *The Supply of Energy to Industry*, op.cit.
35. Exxon Corporation, *The Role of Synthetic Fuels in the US Energy Future* (New York: Exxon Corporation, 1980).
36. Hagler, Bailly, Inc., *Alternative Fuels Monitor* (Washington, D.C.: US Environmental Protection Agency, 1980).
37. S. Eriksson, *Economic and Technical Criteria for Coal Utilization Plant, Part II: Gasification* (London: IEA Coal Research, Economic Assessment Service, 1977).
38. Booz, Allen and Hamilton, Inc., *Analysis of industrial markets for low and medium Btu gas* (Bethesda, Md: Booz, Allen and Hamilton, Inc, 1979).

CHAPTER 3   COAL, ENVIRONMENT AND HEALTH

1. C. G. Downs and J. Stocks, *Environmental Impact of Mining* (London: Applied Science Publishers, 1977).
2. For an overview, see Office of Technology Assessment, *The Direct Use of Coal* (Washington, D. C.: US Government Printing Office, 1979) Ch. V; *Report of the Committee on Health and Environmental Effects of Coal Use* ('Rall Report') (Washington, D. C.: US Environmental Protection Agency, 1978); United Nations Economic Commission for Europe, *Environment and Energy* (London: Pergamon, 1979); C. W. Gehrs *et al.*, 'Environmental health and safety implications of increased coal utilisation', in M. A. Elliott (ed.), *Chemistry of Coal Utilisation* (New York: Wiley, 1980); Commission on Energy and the Environment, *Coal and the Environment* (London: Her Majesty's Stationery Office, 1981).
3. For an overview, see C. G. Downs and J. Stocks, op. cit.; US Department of Energy, *Coal Extraction and Preparation Technology: Environmental Readiness Document* (Washington, D.C.: US Department of Energy, 1979); M. J.

Chadwick and G. T. Goodman, *The Ecology of Resource Degradation and Renewal* (Oxford: Basil Blackwell, 1975).

4. W. S. Doyle, *Deep Coal Mining Waste Disposal Technology* (Park Ridge, NJ: Noyes Data Corporation, 1976).

5. R. B. Dunn, 'Mining waste', *Mining Technology*, 60 (1978) pp. 319–27; Office of Technology Assessment, op. cit. p. 251.

6. Bureau of Mines, *Coal Refuse Fires* (Washington, D. C., US Department of Interior, 1971).

7. C. O. Brawner and I. Dorling (ed.), *Stability in Coal Mining* (San Francisco: Miller-Freeman Publications, 1979); A. D. Bradshaw and M. J. Chadwick, *The Restoration of Land* (Oxford: Basil Blackwell, 1980).

8. W. S. Doyle, *Strip Mining of Coal – Environmental Solutions* (Park Ridge, NJ: Noyes Data Corporation, 1976); World Coal, *Reclamation of Surface Mined Lands* (San Fransisco: Miller-Freeman Publications, 1979), summarised in *World Coal*, 5 (1979) p. 56; A. D. Bradshaw, op. cit; National Research Council, *Coal: Soil, Society and Environment* (Washington DC: National Academy of Sciences, 1981).

9. P. Kausch, 'Brown coal opencast mining', *Energy Research*, 3 (1979) pp. 275–88.

10. Office of Technology Assessment, op. cit. pp. 232–6.

11. US Environmental Protection Agency, *Dewatering Active Underground Mines* (Washington, D. C.: US Environmental Protection Agency, 1979).

12. US Environmental Protection Agency, *Removal of Trace Elements from Acid Mine Drainage* (Washington, D. C.: US Environmental Protection Agency, 1979).

13. US Senate Committee on Governmental Affairs, *The Coal Industry: Problems and Prospects* (Washington, D. C.: US Government Printing Office, 1978) p. 109.

14. Comptroller General of the USA, *Alternatives to Protect Property Owners from Damages Caused by Mine Erosion* (Washington, D.C.: US General Accounting Office, 1979) pp. 3–4; also Institute of Civil Engineering, *Ground Subsidence* (London: Institute of Civil Engineers, 1977).

15. P. Wood, *Less-conventional underground coal mining* (London: IEA Coal Research, Technical Information Service, 1980).

16. US Department of Energy, *Coal Extraction and Preparation Technology*, op. cit. p. 1.

17. National Academy of Sciences, *Critical Issues in Coal Transportation Systems* (Washington, D.C.: National Academy of Sciences, 1979); M. F. Szabo, *Environmental Assessment of Coal Transportation* (Cincinatti: PEDCO Environmental, 1978).

18. R. N. Palmer *et al.*, *Comparative assessment of water use and environmental implications of coal slurry pipelines* (Washington, D.C.: US Geological Survey, 1977).

19. M. W. Miller *et al.*, 'High voltage overhead', *Environment*, 20 (1978) pp. 6–15, 32–6.

20. For an overview, see US Department of Energy, *Environmental Characterisation Information Report: Coal-Fired Power Plant* (Washington, D.C.: US Department of Energy, 1980); Central Electricity Generating Board, *Environmental Impact of Coal Consumed in Electricity Generation* (London:

Central Electricity Generating Board, 1979); G. F. Morrison, *Hot Gas cleanup* (London: IEA Coal Research, Technical Information Service, 1979).

21. M. Prior *et al.*, *Control of Sulphur Oxides Emitted from Coal Combustion* (London: IEA Coal Research, Economic Assessment Service, 1977); N. H. Highton and M. G. Webb, 'Sulphur Dioxide from electricity generation', *Energy Policy* (1980) pp. 61–76.

22. Various articles on 'Flue gas desulferisation', *Chemical Engineering Progress*, 76 (1980) pp. 45–68.

23. National Research Council, *U.S. Energy Supply Prospects to 2010* (Washington, D.C.: National Academy of Sciences, 1980).

24. Central Electricity Generating Board, op. cit. p. 11; OECD, *The Costs and Benefits of Sulphur Oxide Control* (Paris, 1981).

25. National Research Council, *Nitrogen Oxides* (Washington, D. C.: National Academy of Sciences, 1977); J. E. Cichanowicz, 'Controlling oxides of nitrogen', *EPRI Journal*, 4 (1979) pp. 22–7.

26. 'Controlling particulate emissions', *Power*, 124 (1980) pp. S1–S20; National Research Council, *Airborne Particles* (Washington, D.C.: National Academy of Sciences, 1977); 'Upsurge in Baghouse development', *EPRI Journal*, 5 (1980) pp. 15–20.

27. M. Y. Lim, *Trace elements from coal combustion* (London: IEA Coal Research, Technical Information Service, 1979); S. P. Baku (ed.), *Trace elements in fuel* (Washington, D. C.: American Chemical Society, 1975); D. H. Klein *et al.*, Pathways of thirty-seven trace elements through coal-fired power plant', *Environmental Science and Technology*, 9 (1975) pp. 973–9.

28. J. W. Jones *et al.*, *Environmental Management of Effluents and Solid Wastes from Steam Electric Generating Plants* (Washington, D.C.: US Environmental Protection Agency, 1977).

29. Office of Technology Assessment, op. cit. p. 252.

30. J. Reynolds, 'Power plant cooling systems', *Science*, 207 (1980) pp. 367–72; J. Harte and M. El-Gasseir, 'Energy and water', *Science*, 199 (1978) pp. 623–33.

31. S. I. Rapeol (ed.), *Chemistry of the Lower Atmosphere* (New York: Plenum, 1973).

32. R. A. Barnes, 'The long range transport of air pollution', *Journal of the Air Pollution Control Association* (1979) pp. 1219–35; Organisation for Economic Co-operation and Development, *Long Range Transport of Atmospheric Pollutants* (Paris: Organisation for Economic Co-operation and Development, 1977).

33. Office of Technology Assessment, op. cit. p. 199.

34. J. B. Mudd and T. T. Kozlowski, *Responses of Plants to Air Pollutants* (New York: Academic Press, 1975); National Research Council, *Atmosphere–Biosphere Interactions* (Washington, D.C.: National Academy of Sciences, 1981).

35. J. O. Nriagu (ed.), *Sulfur in the Environment* (New York: Wiley, 1978); J. N. Bell, 'Response of plants to sulphur dioxide', *Nature*, 284 (1980) pp. 399–400.

36. National Research Council, *Ozone and Other Photochemical Oxidants* (Washington, D.C.: National Academy of Sciences, 1977).

37. G. E. Likens *et al.*, 'Acid rain', *Scientific American*, 241 (1979) pp. 42–51; H.

Babich *et al.*, 'Acid precipitation: 1', *Environment*, 22 (1980) pp. 6–13; G. S. Whetstone, 'Acid precipitation: 1' and A. Rosencrantz, 'Acid precipitation: 2', *Environment*, 22 (1980) pp. 9–14 and 15–25; T. C. Hutchison and M. Havas (eds), *Effects of Acid Precipitation on Terrestrial Ecosystems* (New York: Plenum, 1980).

38. For an overview, see Climate Research Board, *Carbon Dioxide and Climate* (Washington, D.C.: National Academy of Sciences, 1979); Commission for Atmospheric Sciences, *Climatic Effects of Increased Carbon Dioxide* (Geneva: World Meteorological Organisation, 1979); I. Smith, *Carbon Dioxide and the 'greenhouse' effect* (London: IEA Coal Research, Technical Information Service, 1978); H. W. Bernard, *The Greenhouse Effect* (Cambridge, Ma: Ballinger, 1980); B. Borin (ed.), *Modelling the Global Carbon Cycle* (New York: Wiley, 1980).

39. T. M. Wigley *et al.*, 'Scenario for a warm, high-CO_2 world', *Nature*, 283 (1980) pp. 17–21; W. Kellogg, 'Modelling future climate', *Ambio*, ix (1980) pp. 216–21; S. Manabe and R. T. Wetherald, 'On the distribution of climate change resulting from an increase in CO_2 content of the atmosphere', *Journal of Atmospheric Sciences*, 37 (1980) pp. 99–118.

40. L. D. Hamilton *et al.*, *Comparative analysis of health and environmental effects of coal conversion technologies* (Washington, D.C.: US Department of Energy, 1978); Organisation for Economic Co-operation and Development, *Potential environmental impacts from the production of synthetic fuels from coal* (Paris: Organisation for Economic Co-operation and Development, 1977); S. C. Morris *et al.*, 'Coal conversion technologies: some health and environmental effects', *Science*, 206 (1979) pp. 654–62; US Department of Energy, *Environmental Impact of Synthetic Liquid Fuels* (Washington, D.C.: US Department of Energy, 1979).

41. US Environmental Protection Agency, *Assessment and Control of Wastewater Contaminants Originating from the Production of Synthetic Fuels from Coal* (Washington, D.C.: US Environmental Protection Agency, 1980); S. Eriksson, *Treatment of liquid effluents from coal gasification plants* (London: IEA Coal Research, Economic Assessment Service, 1979).

42. US Department of Energy, *Environmental Readiness of Emerging Energy Technologies* (Washington, D.C.: US Department of Energy, 1980).

43. For an overview, see US Department of Energy, *Comparative Assessment of Health and Safety Impacts of Coal Use* (Washington, D.C.: US Department of Energy, 1980); Health and Safety Commission, *The Hazards of Conventional Sources of Energy* (London: Her Majesty's Stationery Office, 1978).

44. J. D. McAteer, *Coal Mine Health and Safety* (New York: Praeger, 1973); Office of Technology Assessment, op. cit. pp. 259–90.

45. The President's Commission on Coal, *Staff Findings* (Washington, D.C.: US Government Printing Office, 1980) pp. 43–4.

46. International Labour Organisation, *Year Book of Labour Statistics* (Geneva: International Labour Organisation, 1980).

47. 'The hazards to health in the hydrogenation of coal', *Archives of Environmental Health*, 1 (1960) pp. 181–231, various articles'; S. Mazumdar *et. al.*, 'An epidemiological study of coal tar pitch volatiles among coke oven workers', *Journal of the Air Pollution Control Association*, 25 (1975) pp. 382–9.

48. R. Wilson *et al.*, *Health Effects of Fossil Fuel Burning* (Cambridge, Mass.: Ballinger, 1980); L. B. Lave and E. P. Seskin, *Air Pollution and Human Health* (Baltimore, Md: Johns Hopkins University Press, 1977); M. G. Morgan *et al.*, 'A probabilistic methodology for estimating air pollution health effects from coal-fired power plants', *Energy Systems and Policy*, 2 (1978) pp. 287–310.

49. F. P. Perera and A. K. Ahmed, *Respirable Particles* (Cambridge, Ma: Ballinger, 1979); M. Y. Lim, op. cit. pp. 31–41; C. E. Crisp, 'Mutagenicity of filtrates from respirable coal fly ash', *Science*, 199 (1978) pp. 73–5.

50. R. I. Van Hook, 'Potential health and environmental effects of trace elements and radionuclides from increased coal utilisation' (Oak Ridge, Th.: Oak Ridge National Laboratory, 1978); D. F. S. Natusch, 'Potentially carcinogenic species emitted to the atmosphere by fossil-fuelled power plants', *Environmental Health Perspectives*, 22 (1978) pp. 79–90.

51. US Department of Energy, *Comparative Assessment . . .* , op. cit. p. 53.

52. National Research Council, *Nitrogen Oxides*, op. cit.; National Research Council, *Nitrates: an Environmental Assessment* (Washington, D.C.: National Academy of Sciences, 1978).

53. National Research Council, *Energy in Transition 1985–2010* (Washington, D.C.: National Academy of Sciences, 1979) p. 545.

54. Office of Technology Assessment, op. cit. p. 219.

55. National Research Council, *Energy in Transition . . .* op. cit. p. 587.

56. National Research Council, *Risks and Impacts of Alternative Energy Systems* (Washington, D.C.: National Academy of Sciences, 1980); A. V. Cohen and D. K. Pritchard, *Comparative risks of electricity production systems: a critical survey of the literature* (London: Health and Safety Executive, Her Majesty's Stationery Office, 1980); L. Gaines *et al.*, *The Total Social Cost of Coal and Nuclear Power* (Cambridge, Ma: Ballinger, 1980); H. Inhaber, 'Risks from conventional and unconventional energy sources', *Science*, 203 (1979) pp. 718–23, and subsequent correspondence; Various articles on 'Respective risks of different energy systems', *IAEA Bulletin*, 5 (1980) pp. 35–128.

57. H. W. Lorber, 'Small power plants: Health and safety', *Environment*, 22 (1980) pp. 25–31; D. B. Champion and M. D. Williams, 'Small power plants: the environmental effects', *Environment*, 22 (1980) pp. 25–32.

58. International Atomic Energy Agency, *Biological and Environmental Effects of Low-Level Radiation* (Vienna: International Atomic Energy Agency, 1977); 'Low-level radiation', *Ambio*, IX (1980), various articles.

59. US Nuclear Regulatory Commission, *Reactor Safety Study* (Washington, D.C.: US Nuclear Regulatory Commission, 1975) and *Report of the Assessment Review Group to the US Nuclear Regulatory Commission*' (Washington, D.C.: US Nuclear Regulatory Commission, 1978); J. Beyea, *A Study of the Consequences of Hypothetical Reactor Accidents at Barseback* (Stockholm: Industridepartment, 1978).

60. National Research Council, *Energy in Transition . . .* , op. cit. p. 582.

CHAPTER 4   THE USA: SAUDI ARABIA OF COAL?

1. For general accounts of the industry, see Office of Technology Assessment, *Direct Use of Coal* (Washington, D. C.: US Government Printing Office, 1979); National Research Council, *Energy in Transition 1985–2010* (Washington, D. C.: National Academy of Sciences, 1979) Ch. 5; R. H. Quenon, 'United States: coal's renaissance may be at hand', *World Coal*, 6 (1980) pp.50–2; US Energy Information Administration, *Coal Data: A Reference* (Washington, D. C.: US Department of Energy, 1980).
2. US statistics use the short ton (st) as a standard unit, a convention which is followed in this chapter, where mst refers to short tons, mt to metric tonnes.
3. P. Averitt, *Coal Resources of the United States, January 1 1974* (Washington, D. C.: US Geological Survey, 1975).
4. ICF, Inc., *Coal Resource Information* (Palo Alto, Ca: Electric Power Research Institute, 1980).
5. W. H. Miernyk, 'Coal and the future of the Appalachian economy', *Appalachia*, IX (1975) pp. 29–35; US Energy Information Administration, *Demonstrated Reserve Base of Coal in the United States on 1 January 1979* (Washington, D. C.: US Department of Energy, 1981).
6. A. M. Sullivan, 'Anthracite seeks developing markets', *Coal Age*, 85 (1980) pp. 87–92.
7. R. Malhotra, 'Midwest coal', *Coal Mining & Processing*, 17 (1980) pp. 46–76.
8. 'Special report on Western coal', *Coal Age*, 85 (1980) various articles pp. 67–171; W. E. Tyner and R. J. Kalter, *Western Coal: Promise or Problem ?* (Lexington, Ma: Lexington Books, 1978).
9. M. D. Devine *etal.*, 'Energy from the western states of the USA', *Energy Policy*, 8 (1980) pp. 229–34; US Environmental Protection Agency, *Energy from the West* (Washington, D.C.: US Environmental Protection Agency).
10. US Energy Information Administration, *Evaluation of Effects of Alternative Western Freight Rates on Coal* (Washington, D. C.: US Department of Energy, 1980).
11. Exxon Corporation, Inc, *'The Role of Synthetic Fuels in the US Energy Future'* (Houston: Exxon Corporation, 1980).
12. 'West shows solid growth', *Coal Age*, 85 (1980) p. 94.
13. Office of Technology Assessment, op. cit. Ch. VII, pp. 337–72.
14. W. Rosenbaum, *Coal and Crisis: The Political Dilemma of Energy Development* (New York: Praeger, 1978); Policy Planning and Evaluation, Inc., *Impact of Government Regulations on Leadtimes of Coal Facilities'* (Washington D. C.: US Department of Energy, 1980)
15. US Department of Interior, *Federal Coal Management Report 1979* (Washington, D. C.: US Department of Interior, 1980).
16. *Coal Age*, 85 (1980) p. 68.
17. D. Jackson, 'Indians want more for minerals', *Coal Age*, 84 (1979) pp. 104–14.
18. R. E. Bailey, 'Obstacles to greater coal use in the USA', *Coal and Energy Quarterly*, 23 (1979) pp. 10–16.

19. Congressional Research Service, *Federal Government Incentives to Coal and Nuclear Energy* (Washington, D. C.: US Government Printing Office, 1979).
20. National Research Council, op. cit. p. 233.
21. Congressional Research Service, *The Coal Industry: Problems and Prospects* (Washington, D.C.: US Government Printing Office, 1978).
22. US Bureau of Mines, *The State of the US Coal Industry* (Washington, D.C.: US Department of Interior, 1976).
23. US Energy Information Administration, *Economic Analysis of Coal Mining Costs for Underground and Strip Operation* (Washington, D.C.: US Government Printing Office, 1978).
24. Office of Technology Assessment, op. cit. pp. 111–20; National Coal Association, *Implications of Investment in the Coal Industry by Firms from Other Energy Industries* (Washington, D.C.: National Coal Association, 1977).
25. The President's Commission on Coal, *Coal Data Book* (Washington, D.C.: US Government Printing Office, 1980).
26. R. Harrold, 'Oil seeks top role in coal game', *Coal Age*, 84 (1979) pp. 64–8.
27. US Department of Justice, *Competition in the Coal Industry* (Washington, D.C.: US Department of Justice, 1979); Comptroller General of the USA, *The State of Competition in the Coal Industry* (Washington, D.C.: US General Accounting Office, 1977); US Department of Energy, *Coal Competition: Prospects for the 1980s* (Washington, D.C.: US Department of Energy, 1981).
28. 'TVA wants an assured coal supply, but not from big oil', *Coal Age*, 84 (1979) p. 25.
29. National Research Council, *U.S. Energy Supply Prospects to 2010* (Washington, D.C.: National Academy of Sciences, 1979).
30. D. W. Walton and P. W. Kauffman, *Preliminary Analysis of the Probable Causes of Decreased Coal-Mining Productivity* (Reston, Va: Management Engineers, Inc., 1977).
31. The President's Commission on Coal, *The American Coal Miner* (Washington, D.C.: US Government Printing Office, 1980).
32. J. Gaventa, *Power and Powerlessness* (Oxford: Oxford University Press, 1980); Appalachian Regional Commission, *Appalachia – A Reference Book* (Washington, D.C.: Appalachian Regional Commission, 1979); Office of Technology Assessment, op. cit. pp. 290–324; Comptroller General of the USA, *Rocky Mountain Energy Development: Status, Potential and Socioeconomic Issues* (Washington, D.C.: US General Accounting Office, 1977).
33. M. Dubovsky and W. Van Tine, *John L. Lewis* (New York: Quadrangle, 1977).
34. The Conference Board, *The Labour Outlook for the Bituminous Coal Mining Industry* (Palo Alto, Ca: Electric Power Research Institute, 1980).
35. Charles Rivers Associates, Inc., *Coal Price Formation* (Palo Alto, Ca: Electric Power Research Institute, 1977); Department of Justice op. cit.; R. W. Schmidt, 'Coal-based electricity in the USA', *Coal and Energy Quarterly*, 27 (1980) pp. 16–22.
36. World Coal Study, *Future Coal Prospects* (Cambridge, Ma: Ballinger, 1980) Ch.16.

37. International Energy Agency, *Steam Coal: Prospects to 2000* (Paris: Organisation for Economic Co-operation and Development, 1978) pp. 104–8; US Energy Information Administration, *Annual Report to Congress 1979* (Washington, D.C.: US Government Printing Office, 1980).
38. Teknekron, Inc., *Projections of Utility Coal Movement Patterns 1980–2000* (Washington, D.C.: Office of Technology Assessment, 1977).
39. American Iron and Steel Institute, *Evidence to the President's Commission on Coal 30 May 1979* (Washington, D.C.: US Government Printing Office, 1979).
40. Energy Modelling Forum, *Coal in Transition 1980–2000* (Stanford, Ca: Stanford University, 1978).
41. Congressional Budget Office, *Replacing Oil and Gas with Coal: Prospects in the Main Industries* (Washington, D.C.: US Government Printing Office, 1978).
42. Exxon Company, USA, *Energy Outlook 1980–2000* (Houston: Exxon Corporation, 1979).
43. Hagler, Bailly, Inc., *Alternative Fuels Monitor* (Washington, D.C.: US Environmental Protection Agency, 1980).
44. R. Peckham, 'Confronting constraints to steam coal exports', *Bulk Systems*, 2 (1980) pp. 10–27; Interagency Coal Export Task Force, *Interim Report* (Washington, D.C.: US Department of Energy, 1981).
45. World Coal Study, *Coal: Bridge to the Future* (Cambridge, Ma: Ballinger, 1980) p. 115.
46. Congressional Research Service, *The Coal Industry: Problems and Prospects* (Washington, D.C.: US Government Printing Office, 1978).
47. C. O. Brawner and I. Dorling (ed.), *Stability in Coal Mining* (San Francisco: Miller-Freeman Publications, 1979) pp. 430–7; National Academy of Sciences, *The Rehabilitation Potential of Western Coal Lands* (Cambridge, Ma: Ballinger, 1974); National Research Council, *Surface Mining: Soil, Coal and Society* (Washington, D.C.: National Academy of Sciences, 1981).
48. Comptroller General of the USA, *Alternatives to Protect Property Owners from Damages Caused by Mine Subsidence* (Washington, D.C.: US General Accounting Office, 1979).
49. B. A. Ackerman and W. T. Hassler, *Clean Coal/Dirty Air* (Yale University Press, 1981).
50. 'EPA says utility conversion to coal cheaper in long run', *Coal Age*, 84 (1979) p. 15.
51. US Department of Energy, *Coal Extraction and Preparation Technology: Environmental Readiness Document* (Washington, D.C.: US Department of Energy, 1980); C. M. Boris and J. V. Krutilla, *Water Rights and Energy Development in the Northern Great Plains* (Cambridge, Ma: Ballinger, 1979).
52. Energy Modelling Forum, op. cit.
53. National Coal Policy Project, *Where We Agree* (Washington, D.C.: Georgetown University, 1979); T. Alexander, 'A promising try at environmental detente for coal', *Fortune*, 97 (1978) pp. 94–102.
54. US Department of Transportation, *Transporting the Nation's Coal – A Preliminary Assessment* (Washington, D.C.: US Department of Transportation, 1978).

55. World Coal Study, *Future Coal Prospects*, op. cit.; National Academy of Sciences, *Critical Issues in Coal Transportation Systems* (Washington, D.C.: National Academy of Sciences, 1979); D. M. Stacey, *Factors Affecting Future Expansion of the Coal Transportation Network* (Washington, D.C.: US Department of Energy, 1980).
56. US Energy Information Agency, op. cit.
57. US Department of Transportation, op. cit.
58. US Secretary of Transportation, *Highway Needs to Solve Energy Problems* (Washington, D.C.: US Government Printing Office, 1978).
59. J. M. Witten and S. A. Desai, *Water Transportation Requirements for Coal Movement in 1985* (Washington, D.C.: US Department of Transportation, 1978).
60. TAMS, *Mid-American Ports Study* (Washington, D.C.: US Department of Commerce, 1979).
61. Comptroller General of the USA, *Coal Slurry Pipelines* (Washington, D.C.: US General Accounting Office, 1979); Office of Technology Assessment, *A Technological Assessment of Coal Slurry Pipelines* (Washington, D.C.: US Government Printing Office, 1978).
62. World Coal Study, *Future Coal Prospects*, op. cit.
63. A. Sullivan, 'Port traffic threatens export market' *Coal Age*, 85 (1980) pp. 66–84.
64. US Energy Information Agency, *Annual Report to Congress 1979* (Washington, D.C.: US Department of Energy, 1980).
65. US Energy Information Agency, *Annual Report to Congress, 1980* (Washington, D.C.: US Department of Energy, 1981).
66. Energy Modelling Forum, op. cit.
67. Energy Modelling Forum, op. cit.
68. National Research Council, *Energy in Transition*, op. cit. p. 26.

CHAPTER 5   RED COAL

1. For an overview, see US Central Intelligence Agency, *USSR: Coal Industry Problems and Prospects* (Washington, D.C.: US Central Intelligence Agency, 1980); L. Dienes and L. Shabad, *The Soviet Energy System* (Washington, D.C.: Winston, 1979); 'USSR coal', *World Coal*, 3 (1977) various articles pp. 23–68.
2. 'Poor planning, equipment failures threaten Russian output', *Coal Age*, 84 (1979) p. 41.
3. 'Russian waste', *Coal Age*, 84 (1979) p. 36.
4. Y. L. Khudin, 'Soviet coal mining shows progress', *World Coal*, 6 (1980) pp. 64–5.
5. T. Sealey, 'Low-quality machines hit output', *Financial Times*, 30 April 1981.
6. D. Satter, 'Where some miners are more equal than others', *Financial Times*, 9 January 1981.
7. B. Komarov, *The Destruction of Nature in the Soviet Union* (London: Pluto, 1980).

8. 'Concern for Soviet energy problems', *Petroleum Economist*, XLVII (1980) pp. 230–1.
9. L. Dienes, op. cit. p. 14.
10. World Coal Study, *Coal: Bridge to the Future* (Cambridge, Ma: Ballinger, 1980); US Central Intelligence Agency, op. cit.
11. For an overview, see US Central Intelligence Agency, *Chinese Coal Industry: Prospects Over The Next Decade* (Washington, D.C.: US Central Intelligence Agency, 1979); 'China today', *World Coal*, 5 (1979) various articles pp. 26–45; C. Zhao-Ning, 'Coal is China's primary source of energy', *World Coal*, 6 (1980) pp. 48–9; V. Smil, 'Energy development in China', *Energy Policy*, 9 (1981) pp. 113–26; C. Joyce, 'Is China ready for the revolution?', *New Scientist*, 90 (1981) pp. 636–9.
12. Z. Yao, 'China's coal expansion plans', *Coal and Energy Quarterly*, 26 (1980) pp. 10–14.
13. US Central Intelligence Agency, *Chinese Coal Industry*, op. cit. p. iii.
14. C. Seltzer, 'The Chinese still get their coal the hard way', *Guardian* (London) 23 September 1980.
15. T. Walker, 'Japanese aid boosts China's port capacity', *Financial Times*, 3 June 1980.
16. C. McDougall, 'China supplement VI: Energy', *Financial Times*, 1 October 1980.
17. B. A. Rohmer, 'Eastern Europe: peculiarities of a regional crisis', *Petroleum Economist*, XLVII (1980) pp. 101–3; G. Hoffman, 'Energy projections – oil, gas and coal in the USSR and Eastern Europe', *Energy Policy*, 7 (1979) pp. 232–41.
18. For an overview, see M. Swiss, 'Domestic demand for coal is growing in Poland', *World Coal*, 6 (1980) p. 53; World Coal Study, *Future Coal Prospects* (Cambridge, Ma: Ballinger, 1980) Ch. 13.
19. L. H. Szklorski *et al.*, 'Mechanization in Poland's coal mines', *World Coal*, 5 (1979) pp. 19–21.
20. L. Colitt, 'Silesian miners learn how to take a day off', *Financial Times*, 23 January 1981.
21. M. Dickson, 'Upsets in the growth of world coal trade', *Financial Times*, 23 January 1981.
22. D. Henrichson, 'Poland: Coming to grips with pollution', *Ambio*, x (1981) pp. 40–1.
23. B. A. Rohmer, 'Poland: bleak prospects for energy', *Petroleum Economist*, XLVII (1980) pp. 72–3.
24. K. Pavlovski, 'The Bulgarian coal mining industry', *Glückauf*, 108 (1972) pp. 231–8.
25. R. J. M. Wyllie, 'Hungary instigates development', *World Coal*, 5 (1979) pp. 26–9.

CHAPTER 6   THE WORLD COAL TRADE

1. G. Markon, 'World coal trade in 1979', *World Coal*, 6 (1980) pp. 36–41; D. R. T. Waring, 'The world coal import and export trade', *Coal and Energy*

Quarterly, 18 (1978) pp. 9–17; J. M. Wilcox, *International Trade in Coal* (Canberra: Australian Department of Trade, 1980).

2. International Energy Agency, *Steam Coal: Prospects to 2000* (Paris: Organisation for Economic Co-operation and Development, 1978); World Coal Study, *Coal: Bridge to the Future* (Cambridge, Ma: Ballinger, 1980).

3. M. Dickson, 'Upsets in the growth of world coal trade', *Financial Times*, 23 January 1981.

4. R. Dafter and P. Cheesewright, 'The Seven Sisters back coal as the fuel of the future', *Financial Times*, 22 April 1980.

5. National Energy Advisory Committee, *Australia's Energy Resources* (Canberra: Department of National Development, 1979).

6. For an overview, see World Coal Study, *Future Coal Prospects* (Cambridge, Ma: Ballinger, 1980) Ch. 1; L. Lyons, 'Big increase in Australian exports foreseen', *World Coal*, 6 (1980) pp. 58–9; 'Australia: Special Report', *World Coal*, 4 (1978) various articles pp. 35–65; Joint Coal Board of New South Wales, *Black Coal in Australia 1979–80* (Sydney: Joint Coal Board of New South Wales, 1980).

7. J. R. Chadwick, 'Hunter Valley No. 1 Expanding', *World Coal*, 6 (1980) pp. 19–22.

8. P. Cheesewright, 'Australia pushes its coal exports', *Financial Times*, 8 February 1980.

9. 'Coal mining methods in Australia', *World Coal*, 6 (1980) pp. 36–9.

10. P. Newby, 'Australia supplement V: Coal export boom forecast', *Financial Times*, 23 September 1980.

11. Organisation for Economic Co-operation and Development, *Australia: Economic Survey 1980* (Paris: Organisation for Economic Co-operation and Development, 1980).

12. C. Chapman, 'Australia Supplement VI, Export Capacity', *Financial Times*, 19 August 1981.

13. *Report of the Commission into the Coal Resources of South Africa* (Johannesburg: Department of Mines, 1975); R. J. Friedland, 'South African coal – challenge to Petrick findings', *Energy Policy*, 7 (1979) pp. 71–2.

14. For an overview, see Minerals Bureau, *Coal – A Commodity Review* (Johannesburg: Department of Mines, 1981); 'Coal has bright future in South Africa', *World Coal*, 6 (1980) pp. 54–5; P. Holz, 'The South African mining industry', *Glückauf*, 116 (1980) pp. 156–8; A. Granville and R. Silverburg, 'Energy in South Africa', *Coal and Energy Quarterly*, 26 (1980) pp. 24–34.

15. 'Amcoal and South African coal', *World Coal*, 6 (1980) p. 25.

16. A. Granville, op. cit.; P. King, 'Long term trends in South African coal mining', *Glückauf* 16 (1980) p. 351.

17. Q. Peel, 'International contractors chase a lucrative market', *Financial Times*, 21 April 1980.

18. P. James, 'South Africa stokes up with coal conversion', *New Scientist*, 85 (1980) p. 930; 'Sasol pushes licensing of coal process', *Oil and Gas Journal*, 77 (1979) pp. 108–9.

19. P. Doerell, 'Richards Bay', *World Coal*, 6 (1980) pp. 33–6.

20. For an overview, see Department of Energy, *Discussion Paper on Coal* (Ottawa: Department of Energy, 1980); Department of Energy, *Coal Resources and Reserves* (Ottawa, Department of Energy, 1980); 'Canada: Coal's Strong Future', *World Coal*, 7 (1981) various articles, pp. 40–75.
21. 'Canadian output', *Coal Age*, 85 (1980) p. 48.
22. L. Dotto, 'What acid rain does to our land and water', *Canadian Geographic*, 99 (1979) pp. 36–41.
23. Department of Energy, *The National Energy Program, 1980* (Ottawa: Department of Energy, 1980); C. F. W. Diessel, 'Coal mining in New Zealand', *Glückauf*, 114 (1978) pp. 437–43.
24. M. Keeley, 'Spate of new deals points to an energy alliance', *Financial Times*, 31 October 1980.
25. International Energy Agency, *Energy Policies of IEA Countries: 1979 Review* (Paris: International Energy Agency, 1980).
26. I. Dorling, 'New Zealand's Huntly West No. 1 Mine', *World Coal*, 5 (1979) pp. 10–14; International Energy Agency, op. cit.
27. For an overview, see World Coal Study, *Future Coal Prospects* (Cambridge, Ma: Ballinger, 1980) Ch. 11; Y. Eguchi, 'Japanese energy policy', *Coal and Energy Quarterly*, 27 (1980) pp. 30–5; 'Japan supplement VI: Coal', *Financial Times*, 21 July 1980; S. Iki, 'Japan's coal industry', *Glückauf*, 116 (1980) pp. 96–100.
28. International Energy Agency, *Energy Policies of IEA Countries: 1979 Review*, op. cit.

CHAPTER 7   WESTERN EUROPE

1. A. W. Gordon, *Steam Coal and Energy Needs in Europe to 1985* (London: Economist Intelligence Unit, 1978); United Nations Economic Commission for Europe, *Coal: 1985 and Beyond* (London: Pergamon, 1978).
2. Commission of the European Communities, *The Energy Programme of the European Communities* (Brussels: Commission of the European Communities, 1979); Energy Commission, *European Community Coal Policy* (London: Department of Energy, 1978); N. J. D. Lucas, *Energy and the European Communities* (London: Europa Publications, 1977).
3. Association of the Coal Producers of the Economic Community (CEPCEO), *On Prospects for Coal Consumption in the General Industry Sector of the Community Energy Market* (Brussels; CEPCEO, 1980).
4. Commission of the European Communities, *Outlook for the Long-Term Coal Supply and Demand Trend in the Community* (Brussels: Commission of the European Communities, 1980).
5. J. U. Nef, *Rise of the British Coal Industry* (London: Routledge, 1932).
6. For an overview, see World Coal Study, *Future Coal Prospects* (Cambridge, Ma: Ballinger, 1980) Ch. 15; D. Ezra, *Coal and Energy* (London: Benn, 1980); 'UK: coal commitment', *World Coal*, 6 (1980) various articles pp. 34–64; 'Year of achievement in the United Kingdom', *World Coal*, 6 (1980) pp. 60–1; G. Manners, *Coal in Britain* (London: Allen and Unwin, 1981).

7. Commission on Energy and the Environment, *Coal and the Environment* (London: Her Majesty's Stationery Office, 1981).
8. A. Trueman, *The Coalfields of Great Britain* (London: Edwin Arnold, 1954); E. H. Francis, 'British coalfields', *Science Progress*, 66 (1979) pp. 1–23.
9. J. North and D. J. Spooner, 'The geography of the coal industry in the UK in the 1970s', *Geo Journal*, 2 (1978) pp. 255–72.
10. National Coal Board, *Report and Accounts 1979–80* (London: National Coal Board, 1980).
11. H. Rhodes, 'British Coal International', *Coal and Energy Quarterly*, 15 (1977) pp. 29–34.
12. National Coal Board, *Plan for Coal* (London: National Coal Board, 1974).
13. P. Tregelles, 'Automation in UK mines', *Coal Mining and Processing*, 17 (1980) pp. 110–22.
14. M. Dickson, 'Coal's tough break-even target', *Financial Times*, 4 August 1980.
15. B. J. McCormick, *Industrial Relations in the Coal Industry* (London: Macmillan, 1979).
16. *Report of a Court of Inquiry into a Dispute between the NCB and the NUM* (Wilberforce Report) (London: Her Majesty's Stationery Office, 1972) pp. 2–3.
17. The Electricity Supply Industry in England and Wales, *Medium-Term Development Plan 1980–87* (London: Electricity Council, 1980).
18. Department of Energy, *Evidence to the North-East Leicestershire (Belvoir) Planning Inquiry* (London: Department of the Environment, 1980).
19. World Coal Study, op. cit.
20. National Coal Board, *Report and Accounts*, op. cit.
21. Royal Town Planning Institute, *Evidence submitted to the SCENE Coal Study* (London: Royal Town Planning Institute, 1979).
22. G. Manners, 'Alternative strategies for the British coal industry', *Geographical Journal*, 144 (1978) pp. 224–34; *Evidence to the North-East Leicestershire Coalfield Planning Inquiry* (London: Department of the Environment, 1980); C. Robinson and E. Marshall, *What Future for British Coal?* (London: Institute of Economic Affairs, 1981).
23. Department of Energy, op. cit.
24. For an overview, see World Coal Study, op. cit. Ch. 7; 'West Germany: coal's growing contribution', *World Coal*, 6 (1980) various articles pp. 28–60; H. Reintges, 'The German coal industry – between crisis and renaissance', *Glückauf*, 116 (1980) pp. 379–86; 'West Germany plans to use more coal', *World Coal*, 6 (1980) pp. 56–7.
25. H. Kleinheme, 'Opening up new coal deposits', *Glückauf*, 115 (1979) pp. 539–42; H. Rürup, 'Exploration – plenty of coal in the North Ruhr Area', *World Coal*, 6 (1980) pp. 30–3.
26. M. Kuschel, 'Ibbenbüren – world's deepest coal mine', *World Coal*, 6 (1980) pp. 34–7.
27. H. J. Leuschner, 'Rhineland's brown coal mines', *World Coal*, 6 (1980) pp. 42–4.
28. H. Bremshey, 'Rheinbraun resettles villages in the FRG', *World Coal*, 4 (1978) pp. 29–32.

29. J. Fryer, 'Aid for coal top priority in West Germany', *Coal and Energy Quarterly*, 22 (1979) pp. 16–22.
30. F. K. Brassier, 'Advanced technology in longwall mining', *World Coal*, 6 (1980) pp. 38–41; H. Kundel, 'On-face technology in German coalmines in 1979', *Glückauf*, 116 (1980) pp. 379–86.
31. H. Kleinheme, op. cit.
32. 'W. German silicosis', *Coal Age*, 84 (1979) p. 51.
33. J. Fedler, 'Ruhrkohle seeks world giant status', *Coal Age*, 83 (1978) pp. 124–8.
34. 'West German plans project large coal imports before 1995', *Coal Age*, 85 (1980) p. 45.
35. P. Doerell, 'Coal trade', *World Coal*, 6 (1980) p. 45.
36. 'Developments in coal gasification and hydrogenation', *World Coal*, 6 (1980) pp. 49–51; D. Fishlock, 'Germany's plans to enrich the world's coal', *Financial Times*, 22 October 1980.
37. K. Done, 'West Germany supplement VIII: Domestic answers to heavy dependence on imports', *Financial Times*, 27 October 1980.
38. National Coal Board, *The Belgian Coal Industry* (London: National Coal Board, 1973).
39. World Coal Study, op. cit. Ch. 4.
40. World Coal Study, op. cit. Ch. 6; P. Doerell, 'France rediscovers coal', *World Coal*, 6 (1980) p. 35; N. J. D. Lucas, *Energy in France* (London: Europa Publications, 1979).
41. R. Home and C. Carty, *Irish Coal: Its Characteristics and Potential* (Dublin: Irish Geological Survey, 1979).
42. World Coal Study, op. cit. Ch. 10; P. Betts, 'At last a long-term plan for energy', *Financial Times*, 31 March 1980.
43. Ministry of Economic Affairs, *Summary of the Memorandum on Energy Policy: Pt II, The Coal Memorandum* (Den Haag: Ministry of Economic Affairs, 1980); World Coal Study, op. cit.
44. P. Lendvai, 'Austria supplement V: reliance on Comecon for energy supplies', *Financial Times*, 26 January 1981.
45. K. Whitworth, 'Mining coal on top of the world, Part 1 and 2', *World Coal*, 3 (1977) pp. 17–19 and 4 (1978) pp. 16–18.
46. F. Kerstan, 'The position of the Spanish coal mining industry and its effect on Hunosa', *Glückauf*, 116 (1980) pp. 64–5; K. Whitworth, 'Spain's Arino Mine', *World Coal*, 4 (1978) pp. 22–23.
47. National Swedish Board for Energy Source Development, *Kol i Sverige* (Coal in Sweden) (Stockholm: Department of Industry, 1977); World Coal Study, op. cit. Ch. 14.
48. H. Reintges, 'Is coal an alternative energy for Switzerland too?', *Glückauf*, 116 (1980) p. 295.

CHAPTER 8 THE THIRD WORLD

1. World Bank, *Coal Development Potential and Prospects in the Developing Countries* (Washington, D.C.: World Bank, 1979); also, J. Pettigrew, 'A

review of non-traditional mining countries Part 1 and Part 2', *World Coal*, 2 (1976) pp. 24–7 and pp. 38–9.

2. World Bank, op. cit. See also International Institute for Applied Systems, *Energy in a Finite World* (Vienna).

3. World Bank, *Energy in the Developing Countries* (Washington, D.C.: World Bank, 1980). See also International Institute for Applied Systems Analysis.

4. R. Bosson and B. Varon, *The Mining Industry and the Developing Countries* (Oxford: Oxford University Press, 1977); G. Lanning with M. Mueller, *Africa Undermined* (London: Penguin, 1979).

5. World Coal Study, *Future Coal Prospects* (Cambridge, Ma: Ballinger, 1980) Ch. 19–22. All subsequent references to the WOCOL Report refer to this source.

6. World Coal Study, op. cit. All subsequent references to Charbonnages de France refer to this source.

7. World Bank, *Coal Development Potential*, op. cit. All subsequent references to World Bank production forecasts refer to this source.

8. M. Dickson, 'Nigeria supplement XI: Coal', *Financial Times*, 29 September 1980.

9. J. G. Ward, 'Maaba – Zambia's only coal mine', *World Coal*, 5 (1979) pp. 23–5.

10. T. Hawkins, 'Zimbabwe supplement XIV: Energy', *Financial Times*, 22 April 1980; J. R. Chadwick, 'Zimbabwe's Coal reserves', *World Coal*, 7 (1981), pp. 50–1.

11. R. Cowper, 'ASEAN's plans for non-oil power', *Financial Times*, 17 October 1980.

12. R. N. Bose, 'Afghanistan's coal industry', *World Coal*, 5 (1979) pp. 31–3.

13. For an overview, see 'The coal industry of India', *World Coal*, 5 (1979) various articles pp. 20–52; World Coal Study, op. cit. Ch. 8; R. P. Khosla, 'India's coal development plans', *Coal and Energy Quarterly*, 29 (1981), pp. 24–9; D. Dodwell, 'India supplement XIV: Coal reserves are vital', *Financial Times*, 17 March 1980.

14. R. G. Mahendru, 'Reconstructing Jharia coalfield', *World Coal*, 5 (1979) pp. 39–41.

15. D. Housego, 'Violence and gangsters rule in India's coal fields', *Financial Times*, 7 January 1981.

16. B. Locke, 'Attacking acute problems of growth', *Financial Times*, 2 November 1979; D. Dodwell, 'India supplement XI: Acute power shortage costs industry dear', *Financial Times*, 17 March 1980.

17. D. Dodwell, 'India supplement XI: Demands of the economy put pressure on railways', and 'India supplement XII: Import programme puts strain on ports', *Financial Times*, 17 March 1980.

18. P. C. Mahanti, 'India supplement XVI: Ambitious programme to raise production', *Financial Times*, 19 January 1981.

19. R. Cowper, 'Jakarta prepares for the day the oil runs out', *Financial Times*, 22 October 1980.

20. M. Samuelson, 'US guarantees provide a safety net', *Financial Times*, 24 October 1980.

21. 'South Korea's coal mining industry', *World Coal*, 2 (1976) p. 28;

P. Bowring, 'Authorities move to cut dependence on imports', *Financial Times*, 9 June 1980.

22. M. Cetincelik, 'Turkey's coal industry', *World Coal*, 5 (1979) pp. 54–5; A. McDermott, 'Turkey supplement IX: Serious problems on energy supplies', *Financial Times*, 21 January 1980; M. Cetincelik and E. Tuncali, 'The brown coal potential of southwest Turkey', *World Coal*, 7 (1981) pp. 24–5.
23. K. Fuad, 'Latin America supplement V: Fuel resources on a vast scale', *Financial Times*, 30 June 1980.
24. R. M. Barwick, 'YCF and the coal situation in Argentina', *Glückauf*, 115 (1979) pp. 280–1.
25. E. U. Reutler, 'The Brazilian coal mining industry', *Glückauf*, 109 (1973) pp. 683–90.
26. P. Dyson and I. Dorling, 'Colombia's coal mining industry, part 1 and 2', *World Coal*, 5 (1979) pp. 14–19 and pp. 49–53.
27. H. O'Shaugnessy, 'Colombia supplement III: Giant coal project goes ahead with foreign investment', *Financial Times*, 6 October 1980.
28. M. Arble, 'Mexico plans coal expansion', *Coal Age*, 84 (1979) pp. 146–53; W. Chislett, 'Mexico's energy plans', *Financial Times*, 2 May 1980.
29. 'Major effort to boost Chihuahua's mining production', *Financial Times*, 28 August 1980.
30. K. Brandi, 'Venezuela looks to the future with coal', *World Coal*, 4 (1978) pp. 28–30.

CHAPTER 9 THE FUTURE OF COAL

1. R. Stobaugh and D. Yergin (ed.), *'Energy Future'* (New York: Random House, 1979) p. 10.
2. World Coal Study, *Coal: Bridge to the Future* (Cambridge, Ma: Ballinger, 1980); World Coal Study, *Future Coal Prospects* (Cambridge, Ma: Ballinger, 1980).
3. World Coal Study, *Coal: Bridge to the Future*, op. cit. p. xvi.
4. H. Aoki, 'International co-operation in world coal development', in *Survey of Energy Resources* (London: World Energy Conference, 1980) p. 469.
5. L. Landsberg (ed.), *Energy: The Next Twenty Years* (Cambridge, Ma: Ballinger, 1979) p. 111.
6. Council on Environmental Quality, *The Global 2000 Report to the President* (Washington, D. C.: US Government Printing Office, 1980) p. 1.
7. G. Goodman *et al.*, *The European Transition from Oil* (London: Academic Press, 1981).
8. J. Surrey, 'What has WOCOL done?', *Coal and Energy Quarterly*, 26 (1980) p. 20.
9. R. Long, *Inflation and the Real Cost of Energy* (London: IEA Coal Research, Economic Assessment Service, 1980) p. 32.
10. World Coal Study, *Coal: Bridge to the Future* op. cit. p. xvii.

POSTSCRIPT

1. Chadwick and N. Lindman, *Environmental Implications of Expanded Coal Utilisation* (Oxford: Pergamon, 1982).
   Coal Industry Advisory Board, *Coal Use and the Environment* (Paris: Organisation for Economic Co-operation and Development, 1982).
   W. Clarke (ed.), *Carbon Dioxide Review: 1982* (Oxford: Oxford University Press, 1982).
   IEA Coal Research, *Carbon dioxide – emission and effects* (London: 1982).
2. R. Samples, 'United States', *World Coal*, 8, 6 pp. 89–91.
   R. Gordon, 'prospects for US coal', *Energy Policy*, 9, 4 pp. 279–88.
3. '$770 million synfuel loss projected', *Financial Times*, 12/4/1983.
4. M. Swiss, 'Soviets Welome Advanced Coal Mining Technology', *World Coal*, 8, 6 pp. 98–9.
   J. Stein, 'Soviet energy', *Energy Policy*, 9, 4 pp. 301–15.
5. 'Developments in China', *World Coal*, 8, 6 pp. 92–3.
   L. Hua, 'Energy conservation in China', *Coal and Energy Quarterly*, Spring 1983 pp. 30–1.
6. M. Swiss, 'Poland', *World Coal*, 8, 6 p. 118.
7. S. Lyons, 'Many changes in the Australian Coal Industry', *World Coal*, 8, 6 pp. 100–2.
   S. Harris and T. Ikuta (eds), *Australia, Japan and the Energy Coal Trade* (Canberra: Australia-Japan Research Centre, 1982).
8. 'Coal Could Displace Gold', *World Coal*, 8, 6 pp. 106–7.
9. J. Aylsworth, 'Canada Plans to Double Exports', *World Coal*, 8, 6 pp. 116–17.
   G. Page, 'Canadian coal forges ahead', *Coal and Energy Quarterly*, Summer 1982 pp. 10–18.
10. Monopolies and Mergers Commission, *National Coal Board* (London: HMSO, 1983).
11. H. Bund, 'west German Coal Industry', *World Coal*, 8, 6 pp. 108–9.
12. 'France: Innovation and Technology', *World Coal*, 8, 4 various articles pp. 26–66.
13. R. Dafter, 'Slow growth in use of coal predicted', *Financial Times*, 8/9/1982.
14. Central Electricity Generating Board, *Statement of Case to the Sizewell Public Inquiry* (London: 1982).
15. R. Long, *Constraints on international trade in coal* (London: IEA Coal Research, 1982).
   Coal Advisory Board, *The Use of Coal in Industry* (Paris: Organisation for Economic Co-operation and Development, 1982).
   L. Grainger and J. Gibson, *Coal Utilisation: Technology, Economics and Policy'* (London: Graham and Trotman, 1981) – also an excellent introduction to coal matters in general.
   International Energy Agency, *Coal prospects and policies in IEA countries* Paris: OECD, 1982).
   US Energy Information Administrations, *Prospects for future world coal trade* (Washington D.C., 1982).

Index